Peter D Riley and Mike Cole

chec

checkp●int
Science

Revision Guide

For the Cambridge Secondary 1 Test

HODDER
EDUCATION
AN HACHETTE UK COMPANY

To David and Hazel

Answers to selected questions can be found at www.hoddereducation.com/checkpointextras

This text has not been through the Cambridge endorsement process.

The Publishers would like to thank the following for permission to reproduce copyright material: **p4** © Yio - Fotolia; **p36** *left* © So-Shan Au, *right* © volgariver - Fotolia, *centre* © peteri - Fotolia; **p53** © Andrew Lambert Photography/Science Photo Library; **p54** © Martyn F. Chillmaid; **p88** © Andrew Lambert Photography/Science Photo Library.

Every effort has been made to trace all copyright holders, but if any have been inadvertently overlooked the Publishers will be pleased to make the necessary arrangements at the first opportunity.

Although every effort has been made to ensure that website addresses are correct at time of going to press, Hodder Education cannot be held responsible for the content of any website mentioned in this book. It is sometimes possible to find a relocated web page by typing in the address of the home page for a website in the URL window of your browser.

Hachette UK's policy is to use papers that are natural, renewable and recyclable products and made from wood grown in well-managed forests and other controlled sources. The logging and manufacturing processes are expected to conform to the environmental regulations of the country of origin.

Orders: please contact Hachette UK Distribution, Hely Hutchinson Centre, Milton Road, Didcot, Oxfordshire, OX11 7HH. Telephone: +44 (0)1235 827827. Email education@hachette.co.uk Lines are open from 9 a.m. to 5 p.m., Monday to Friday. You can also order through our website: www.hoddereducation.com.

'Preparing for the test', 'Tips for success' and 'Spotlight on the test' sections written by Mike Cole. © Mike Cole 2013

© Peter D Riley 2013
First published in 2013 by
Hodder Education
An Hachette UK Company
London NW1 3BH

Impression number   9
Year   2023 2022 2021

Illustrations by Barking Dog Art and Gray Publishing
Typeset in 12/14pt Garamond and produced by Gray Publishing, Tunbridge Wells
Printed in India

A catalogue record for this title is available from the British Library

ISBN 978 1 4441 8073 2

Contents

Introduction

Preparing for the test

You might find the following points helpful when preparing for the Cambridge Checkpoint Science Secondary 1 Test.

- Make sure that you are familiar with the science content that the test will cover. Use this revision guide to check your understanding of each topic area. You can tick off each section once you have practised it in the box provided.
- Read the **Tips for success** boxes for useful pointers to the areas of understanding the test is looking to check.
- Make sure that you are familiar with the types of questions that will be asked in the test.
- Try to get hold of some past papers, and practise working through these in the time limits given, so that you know what is expected.
- Spend some time practising the test-style questions (**Spotlight on the test**) in this revision guide.
- Remember to look at the mark scheme so that you know how much each answer is worth.
- Check your answers against the sample answers so that you can see how your answers will be assessed.

General revision tips

- Find somewhere quiet to revise. Sit on a comfortable chair at a table and have a pen or pencil and some sheets of paper as well as this book. Plan what you will revise in your revision session. Remember to take a break; perhaps every 20 or 30 minutes to let your mind rest.
- Just reading through the text is not always the best way of learning. It is better to make your revision more active. You should use a variety of active ways to make your learning secure. Here are some of them:
  - Cover a diagram or table with your hand and try to recall what is there.
  - Read through a section of text and write down each key word (those in bold type) on a piece of paper. Close your book and make a definition of each word or phrase.
  - Read about a process and write down the words that you think are the key words to remember. Then, make a mind map by joining the words together with lines and writing a sentence on each line about how and why the words are linked together.
  - Continue your active revision by completing the **Check your understanding** sections. Write your answers in the book or on some paper.
  - Check your revision progress by responding on paper to each **Tips for success**.

Questions in the **Check your understanding** sections that test your science enquiry skills are identified with a magnifying glass icon and a green background.

☑ Introducing plants

The organs of a plant

A _____

B _____

C _____

D _____

Figure 1.1 The main organs of a plant

- The **root** holds the plant in the soil, and takes up water and minerals from the soil.
- The **stem** supports the leaves and flowers, and transports water and food through the plant.
- The **leaf** produces food by photosynthesis (see page 3).
- The **flower** contains the reproductive organs (see page 6).

Absorption of water and minerals

Water in the soil contains dissolved minerals (see below). The taking up of water and the minerals dissolved in the water is called **absorption**. It takes place in the root hair part of the root.

Transport of water and minerals

- The water and minerals are drawn up the root and stem and into the leaves in tiny tubes called **xylem vessels**.
- Most of the water in the leaves evaporates to form water vapour in the spaces between the leaf cells. The water vapour escapes into the air through holes in the leaf surface called **stomata**.
- This process by which plants lose water is called **transpiration** and the movement of water through the plant is called the **transpiration stream**.

The importance of water to plants

- The chemical reactions of life such as respiration and photosynthesis take place in water. Without water, plants would die.
- A lack of water makes a plant **wilt**, so that the leaves are not held up to the sunlight. Photosynthesis is reduced and then stops.

The importance of minerals to plants

Nitrogen, phosphorus and potassium are three important minerals.

Mineral	Needed for …	Absence causes …
nitrogen	leaf development	yellow leaves and poor growth
phosphorus	root development	poor growth
potassium	flower and fruit development	yellow leaves which grow abnormally

Tips for success

Make sure you can label the parts of a plant, and that you know which minerals are needed for the growth of each part.

Check your understanding

1 Label A–D on Figure 1.1 on page 1.
2 Where does the absorption of water take place?
3 Where does the evaporation of water take place?
4 Draw an arrow on Figure 1.1 to show the transpiration stream. Label the arrow.
5 Which mineral is needed for the growth of A, B and D?

 Spotlight on the test

David is growing a plant on his windowsill. He notices that it has weak shoots and roots. Give a reason for this. [1]

Photosynthesis

Definition of photosynthesis

Photosynthesis is the process by which plants make biomass and oxygen from water and carbon dioxide, using energy in light that has been trapped in chlorophyll.

The stages in photosynthesis

Tips for success

Remember that photosynthesis takes place in the chloroplasts of plant cells.

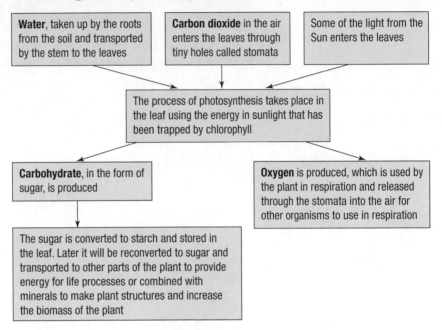

Figure 1.2 The stages in photosynthesis

The starch test

The starch test can be used in three investigations about photosynthesis.

This is how you carry out the starch test.

1 A leaf is dipped in boiling water to soften it.

2 The leaf is then boiled in ethanol to remove the chlorophyll. The leaf is again dipped in boiling water to soften it.

3 Iodine solution is poured over the leaf.

4 A black colour indicates the presence of starch; a brown colour indicates the absence of starch.

A destarched plant

Photosynthesis investigations begin by destarching the plants that will be used. A plant is destarched by putting it in a dark place for two or three days.

Photosynthesis investigations

If a plant does not receive water, it cannot photosynthesise and produce starch. It will eventually die. Three further investigations can be made to study photosynthesis.

1 Carbon dioxide and starch production

Two destarched plants, A and B, are placed in a sunny place for a few hours, as shown in Figure 1.3. The leaves are then tested for starch.

- Soda lime absorbs carbon dioxide from the air, so destarched plant A does not make starch.
- Sodium hydrogencarbonate releases carbon dioxide into the air, so destarched plant B can make starch.

2 Light and starch production

A destarched plant is set up as shown in Figure 1.4 and placed in a sunny place for a few hours. Leaves 1 and 2 are then tested for starch.

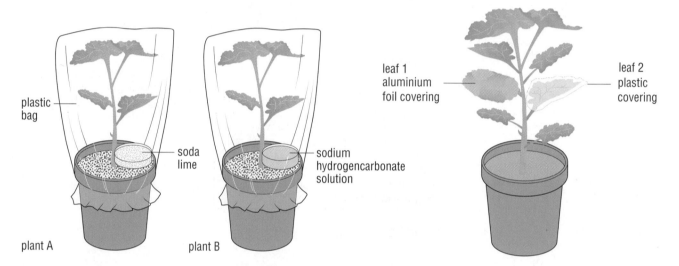

Figure 1.3 Investigating the effect of carbon dioxide on starch production

Figure 1.4 Investigating the effect of light on starch production

3 The effect of chlorophyll on starch production

A **variegated** leaf has green parts where chlorophyll is present, and white parts where chlorophyll is absent.

A destarched plant with variegated leaves is left in a sunny place for a few hours. The leaves are then tested for starch.

Figure 1.5 A variegated leaf

Check your understanding

6 In Figure 1.3, which plant produces starch: A or B?

7 a) In Figure 1.4, which leaf does not produce starch?
   b) Explain your answer.

8 On Figure 1.5, shade in the parts of the variegated leaf that turned black in the starch test.

The production of oxygen

The role of light in photosynthesis can be investigated by setting up two samples of water plants and testing for the production of oxygen (Figure 1.6).

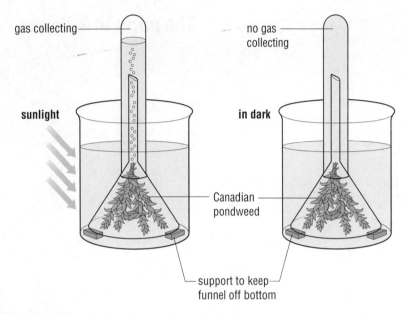

Figure 1.6 The production of oxygen by water plants

Tips for success

Remember that oxygen will relight a glowing splint, but carbon dioxide will extinguish a lit splint.

The water plants are left for a week and the amount of gas in each test tube is observed. Only the plant placed in the light produces gas. The gas is tested for oxygen by plunging a glowing splint into the test tube. If oxygen is present the splint relights.

The word equation for photosynthesis

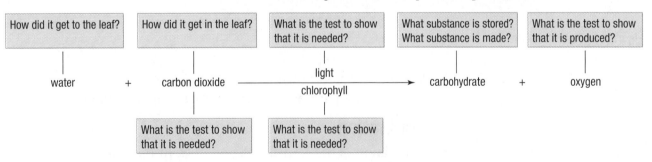

Figure 1.7 All about photosynthesis

Check your understanding

9 On a sheet of paper, make a large copy of Figure 1.7 and complete the answers to each of the questions.

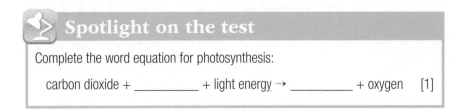

Spotlight on the test

Complete the word equation for photosynthesis:

carbon dioxide + _____ + light energy → _____ + oxygen   [1]

☑ Plant reproduction

The flowers of a plant contain the organs of reproduction. There are two kinds of flowers: insect-pollinated flowers and wind-pollinated flowers.

The parts of a flower

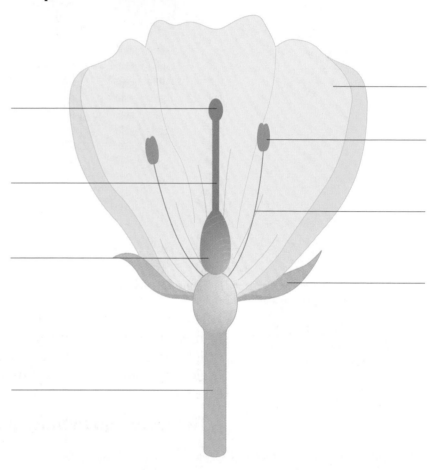

Figure 1.8 Parts of a flower

Tips for success

Make sure you know the structural differences between wind-pollinated flowers and insect-pollinated flowers.

Wind-pollinated flowers do not have scent, may not have petals, and their anthers and stigmas hang outside the flower.

Pollination

Pollination is the transfer of pollen from the **anther** to the **stigma**. After pollination a **pollen tube** grows from the stigma down the style and into the ovary.

Fertilisation

The **male gamete** travels down the pollen tube to the **female gamete** in an ovule. Fertilisation occurs when the male gamete and female gamete fuse.

After fertilisation

The **ovule** forms a **seed** and the **ovary** becomes a **fruit**.

Dispersal of fruit

Plants have ways to disperse fruit to spread out the seeds. This allows the seeds to have enough soil, space and light in which to grow.

Check your understanding

10 Label Figure 1.8 on page 6.
11 Write the following stages of plant reproduction in the order in which they occur:

A fertilisation occurs
B petal attracts insect
C fruits dispersed
D pollen tube grows
E insect feeds on nectar in flower and picks up pollen

F seeds grow
G insect visits another flower and leaves pollen on stigma
H seed and fruit form
I male gamete travels down pollen tube

Spotlight on the test

Look at the pictures below of a dandelion fruit and a maple fruit.

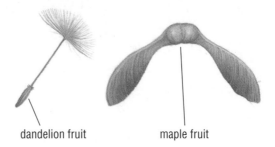

dandelion fruit   maple fruit

Figure 1.9

a) How are they dispersed?   [1]
b) Name *one* feature that helps them to be dispersed in this way.   [1]

☑ Parts of the body

Major organ systems

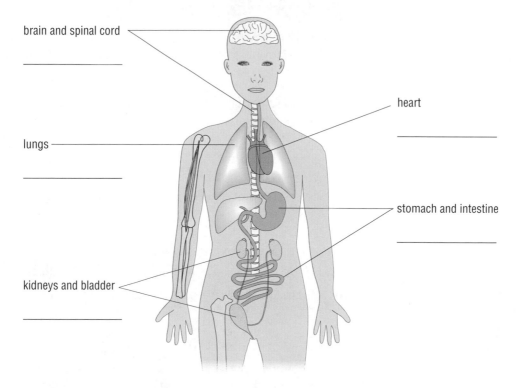

brain and spinal cord

_____

lungs

_____

kidneys and bladder

_____

heart

_____

stomach and intestine

_____

Figure 2.1 The major organs of the body

The organs of the body perform a range of tasks to keep the body alive.

- The **nervous system** uses messages from the sensory organs and sends messages to the muscles to coordinate the activities of the body.
- The **digestive system** breaks down food so that the energy and materials it contains can be used by the body.
- The **respiratory system** takes in oxygen from the air and releases carbon dioxide.
- The **excretory system** removes poisonous liquid wastes.

The skeleton

The skeleton has three functions.

- Protection: the skull protects the brain; the vertebrae, which form the backbone, protect the spinal column; the backbone and ribs protect the heart and lungs.
- Support: the bones provide a strong structure which holds the organs up so they do not squash one another.

● Movement: the place where two bones meet is called a joint. Many joints allow movement. Hinge joints (for example, elbow) allow backwards and forwards movement. Ball and socket joints (for example, hip and shoulder) allow side-to-side movement, too.

At a movable joint you find:
● **ligaments**, which hold the bones together
● **cartilage**, which protects the ends of bones from wear
● **synovial fluid**, which reduces friction as the bones move.

Antagonistic muscles

● A muscle can move. It **contracts** (gets shorter) when it moves. A muscle cannot make itself longer again. It needs another muscle to pull it back to its original length.
● Many muscles are arranged in pairs so that when one muscle contracts it makes the other muscle in the pair relax and lengthen. The action of one muscle is antagonistic or opposite to the other. Muscles arranged like this are called an **antagonistic muscle pair**.

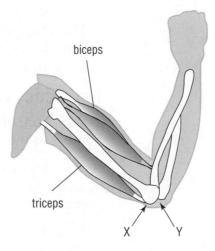

Figure 2.2 Antagonistic muscles

Check your understanding

1 On Figure 2.1 on page 8 write the name of the system to which each organ belongs.
2 Look at Figure 2.2 above.
   a) Which muscle is i) contracted, ii) relaxed?
   b) Draw an arrow on the diagram to show the way the lower arm moves when the muscles swap their actions.
   c) What coats the end of bone X to protect it from wear?
   d) What holds bones X and Y together?

Spotlight on the test

Which antagonistic pair of muscles controls the elbow joint in humans?   [1]

Diet and nutrition

Nutrients

- A **nutrient** is a chemical which is needed by the body to keep it in good health.

Nutrient	Function	Foods rich in the nutrient
protein	growth and repair of body tissues	meat, beans, lentils, milk, cheese
fats	provide energy that is stored in the body	cheese, butter, peanuts
carbohydrates	provide energy which can be used quickly	bread, rice, pasta
vitamins	wide range of functions in keeping the body healthy	fruit, vegetables
minerals	wide range of functions in keeping the body healthy (for example, iron needed for blood to carry oxygen) and maintaining its structure (for example, calcium needed for bones and teeth)	fruit, vegetables

- The body needs water as all the chemical reactions of life processes in the cells take place in water.
- The body needs fibre too. Fibre gives bulk to the food so that the alimentary canal can push it along more easily and make digestion more efficient.

Deficiency diseases

A deficiency disease is a disease caused by the lack of a nutrient in the diet.

Deficiency disease	Lack of nutrient	Foods for prevention of disease
night blindness – cannot see in dark	vitamin A	milk, liver, carrot
beriberi	vitamin B	bread, milk, brown rice, soya bean
scurvy	vitamin C	blackcurrant, orange, lemon, papaya, guava
rickets	vitamin D	egg yolk, butter, pilchards
anaemia	iron	meat, eggs, lentils

A balanced diet

A **balanced diet** is one in which all the nutrients are present in the correct amounts to keep the body healthy.

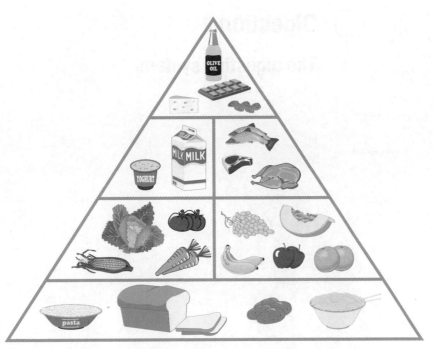

Figure 2.3 Pyramid of food showing the amounts of different foods that can provide a balanced diet

Christiaan Eijkman and beriberi

● Eijkman tried to find out the cause of beriberi. His study of other scientists' research showed that microorganisms cause a range of diseases. He predicted that microorganisms also caused beriberi. However, his investigations did not show this.

● Eijkman observed chickens that showed signs of beriberi. He discovered that their diet had been changed from chicken feed to polished rice. His patients also ate polished rice. When he fed the chickens on chicken feed again they recovered.

● Later work by others showed that rice has a skin containing vitamin B. Polished rice has this skin removed, so the patients were getting a diet deficient in vitamin B and this deficiency caused their disease.

Check your understanding

3 What nutrient would be lacking from the diet of a person who did not grow well?

4 What nutrient would you expect to find in excess in the diet of an obese person?

5 Look at Figure 2.3.
  a) What foods can you eat in large amounts?
  b) What foods should you eat in very small amounts?

6 What previous knowledge did Eijkman use for his first investigations?

7 What evidence did Eijkman collect from first-hand experience?

8 What scientific knowledge and understanding was later used to explain Eijkman's results?

Spotlight on the test

Emma is a body builder and is training hard to develop larger and stronger muscles. Which type of nutrient is particularly important in muscle growth? [1]

Digestion

The digestive system

A _____

B _____

D _____

E _____

F _____

H _____

C _____

G _____

I _____

J _____

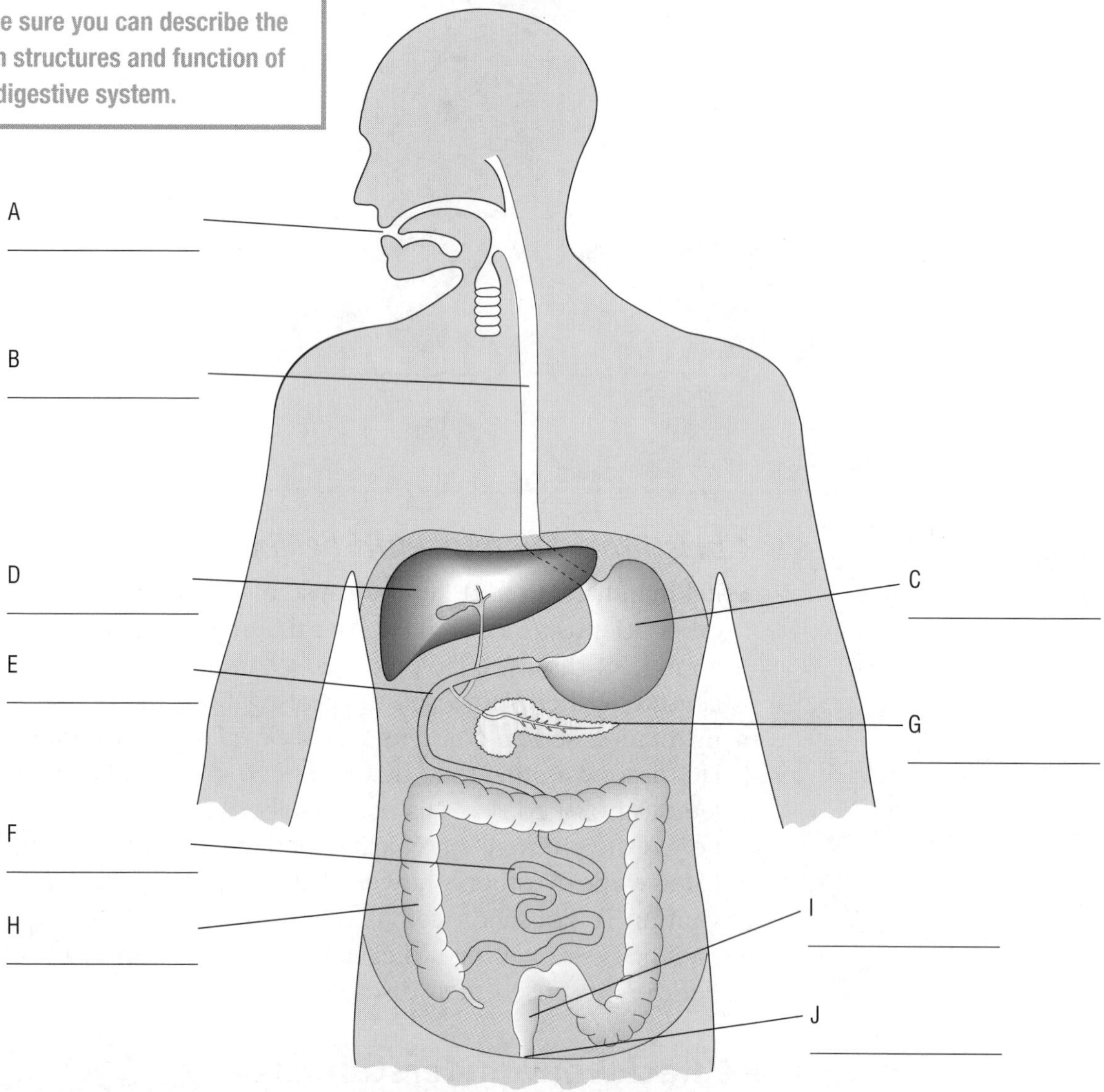

Figure 2.4 The digestive system is made up of the alimentary canal, the liver and the pancreas

Breaking down food

- Food is broken down physically by the action of the teeth and enzymes.
- Food is broken down chemically by the action of enzymes. They act as catalysts and speed up the break down of large molecules in food into smaller ones which can then be absorbed by the body.

Food group	Small molecules produced
protein	amino acids
fats	fatty acids and glycerol
carbohydrates	simple sugars

Studying the stomach

A doctor called William Beaumont treated a man who had been accidentally shot. The man survived but a hole that did not heal was left in his stomach. The patient allowed Dr Beaumont to place food directly into his stomach through the hole and then observe what happened to it. He found that cooked meat broke down faster then uncooked meat.

Check your understanding

9 Match each type of tooth to its function.

| biting | grinding | tearing |

Tooth	Function
incisor	
canine	
molar	

10 On Figure 2.4 on page 12, label the organs A–J.
11 Next to each of your labels on Figure 2.4 write:
   a) where water is reabsorbed
   b) where acid kills bacteria and pepsin starts protein digestion
   c) where undigested food is stored
   d) where food is absorbed
   e) where bile is made
   f) where enzymes for digesting proteins, fats and carbohydrates are made.

12 Did Beaumont use first-hand experience or secondary sources in his investigation?

13 What conclusion about the digestion of meat do you think Beaumont made?

Spotlight on the test

Complete the following sentences:
1 Proteins are broken down by enzymes into _____. [1]
2 Fats are broken down by enzymes into _____. [1]

☑ The circulatory system

The basic parts of the circulatory system

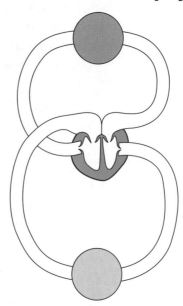

Figure 2.5 A simple diagram of the circulatory system

- The heart pumps the blood around the body.
- Blood is carried away from the heart in arteries. Arteries have thick elastic walls to stand up to the high pressure of the blood as it leaves the heart.
- In the body tissues, the arteries divide into tiny blood vessels called capillaries. Exchange of materials between the blood and the tissue cells takes place in the capillaries.
- The blood moves back towards the heart in veins. Veins have thinner walls as the blood pressure is lower. Veins have valves to prevent the blood flowing backwards.

Components of blood

Component	Function
red blood cells (biconcave disc, no nucleus)	contain haemoglobin which carries oxygen from the lungs to the tissues
white blood cells (irregular shape, has a nucleus)	kill bacteria which cause disease
platelets (fragments of cells)	help blood to clot at the site of a wound
plasma (yellow watery liquid), makes up about 55% of the blood	carries digested food and waste products such as carbon dioxide and urea

The action of the heart

- A simple diagram of the heart is shown in Figure 2.5.
- Blood enters the two **atria** (plural of *atrium*) at the same time. The walls of the atria push the blood through the **valves** into the ventricles.
- The valves close and stop the blood flowing backwards. The walls of the two **ventricles** push the blood out of the heart.

Tips for success

Make sure you can describe and explain the differences in structure between arteries and veins.

The pulse

- As the heart pumps the blood, it makes the blood surge through the blood vessels leaving the heart and causes their walls to stretch and shrink.
- This change in the blood vessel walls can be detected as a throbbing sensation called the **pulse**. The rate of throbbing of the pulse matches the rate of beating of the heart.

Harvey and the heart

William Harvey was taught by a professor who had discovered that veins had valves in them. When Harvey became a doctor, the professor's discovery gave him an idea for an investigation. He thought the valve would only let blood go one way.

He tied a cord around a vein and saw that the blood did not flow 'backwards' through the valve. It could only go one way.

Check your understanding

14 Look at Figure 2.5 on page 14.
  a) Draw arrows to show the movement of blood around the circulatory system.
  b) Colour the blood vessels and the side of the heart which transports oxygenated blood in red. Colour the blood vessels and the side of the heart which transports deoxygenated blood in blue.
  c) Label the blood vessels: A for artery and V for vein. Label where the capillaries are found with C.

15 Did Harvey use first-hand experience or secondary sources for his idea for an investigation?

16 Predict what Harvey saw when he tied off an artery – did the blood flow back to the heart or did it stay in the artery and cause it to swell up?

 Spotlight on the test

Abi runs a 100 metre race.
a) Predict what is likely to happen to her pulse rate. [1]
b) Explain why you made your prediction. [1]

The respiratory system

The function of the respiratory system is to exchange oxygen and carbon dioxide at a rate which meets the needs of the body whether it is active or at rest.

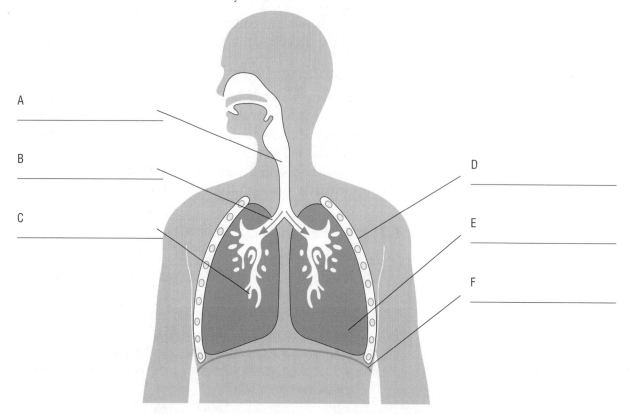

Figure 2.6 The respiratory system

Components of the respiratory system

Component	Features and functions
nose	hairs filter dust; mucus traps bacteria and moistens air for lungs; blood vessels in the lining warm air for lungs
windpipe (trachea)	cartilage rings hold airway open all the time; ciliated cells move dust and bacteria trapped in mucus to top of windpipe for swallowing
lung	organ where gaseous exchange takes place; has three components
bronchus	tube to each lung with same features as the windpipe
bronchioles	tubes 1 mm in diameter with muscles in walls carry air to alveoli
alveoli (singular alveolus)	tiny bubble-like structures with very thin walls, one cell thick, which form a huge surface through which gaseous exchange takes place
chest wall	ribs and muscles; ribs rise to increase chest volume and draw air in, ribs lower to decrease chest volume and push air out
diaphragm	sheet of muscle at base of chest; contracts and lowers to increase chest volume and draw air in, relaxes and rises to decrease chest volume and push air out

Tips for success

Make sure you can explain why the process of respiration is so vital.

Gaseous exchange

- The **alveoli** form the respiratory surface through which gaseous exchange takes place.
- Oxygen from the inhaled air dissolves in the moist covering of the alveolar wall, then diffuses through the wall into the capillary next to it and is taken up by haemoglobin in the red cells to form **oxyhaemoglobin**.
- Carbon dioxide dissolved in the plasma diffuses from the capillary through the alveolar wall, then escapes as a gas into the lung air and is exhaled.

Aerobic respiration

In aerobic respiration, energy is released from food with the use of oxygen. Aerobic respiration takes place in the body cells. The word equation for this reaction is shown in Figure 2.7.

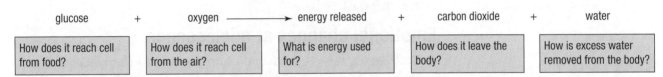

glucose   +   oxygen ⟶ energy released   +   carbon dioxide   +   water

| How does it reach cell from food? | How does it reach cell from the air? | What is energy used for? | How does it leave the body? | How is excess water removed from the body? |

Figure 2.7 Aerobic respiration

Check your understanding

17 On Figure 2.6 on page 16, label the parts of the diagram A–F.

18 What two things happen for:
   a) air to be drawn into the lungs?
   b) air to be pushed out of the lungs?

19 On a sheet of paper make a large copy of Figure 2.7. Complete the answers to each question.

Spotlight on the test

Complete the word equation for respiration.

_____ + oxygen → _____ + water (+ energy)   [1]

Human reproduction

- Humans become capable of reproducing when they have gone through the period of physical changes called **puberty** during adolescence.
- **Adolescence** is the time from the beginning of puberty to the end of the emotional changes, after which a person becomes an **adult**.

Physical changes at puberty

Physical changes are brought about by the action of the growth hormones and sex hormones.
- Males: growth of hair on face, in the armpits and in the pubic region; broadening of shoulders; growth of penis and testicles; deepening of voice.
- Females: growth of hair on face, in the armpits and in pubic region; development of breasts; growth of vagina and uterus; widening of pelvis.

Emotional changes at adolescence

- Both sexes have emotional changes due to their physical changes and a desire to be more independent.
- Anxiety can arise due to: the variation between people in the rate and timing of the physical changes; variations in the ways different families allow independence; increasing interest in the opposite sex due to the action of sex hormones; fear of losing friends due to peer pressure in which others encourage a person to follow their ways.

The reproductive systems

Reproductive system – organs	Features and function
testis (two testicles) ♂	contained in a sac called the scrotum to keep them cool for sperm production
glands along tube to the outside, called the urethra ♂	glands produce fluid which mixes with sperm to form semen
penis ♂	delivers semen into vagina
ovary (two ovaries) ♀	in the lower torso, produce eggs
oviduct ♀	one next to each ovary, contains ciliated cells which move egg along the oviduct
uterus ♀	connected to each oviduct; place where foetus develops
vagina ♀	receives penis and semen

The menstrual cycle

- The wall of the uterus thickens every month in case a fertilised egg is received.
- If no fertilised egg is received, the wall breaks down, causing bleeding from the vagina.

Fertilisation

- The head of the sperm cell contains the nucleus, which is the male gamete. The nucleus of the egg is the female gamete.
- Sperm swim from the vagina through the uterus into the oviducts. If an egg is present in an oviduct the head of a sperm breaks off and enters the egg.
- Fertilisation occurs when the male and female gametes join together and form a cell called the **zygote**.

Foetal development

- The zygote divides to form a ball of cells, which implants in the uterus wall. The cells on the surface of the zygote form the **placenta**. The placenta is the organ which takes food and oxygen to the foetus from the mother's blood and takes wastes such as carbon dioxide from the foetus back to the mother's blood.
- The placenta produces hormones which stop the ovaries producing more eggs and stop the uterus wall breaking down. Inside the ball of cells the foetus forms and grows.
- The foetus is connected to the placenta by the **umbilical cord** and is surrounded by a bag called the **amnion**, which is filled with a watery liquid called **amniotic fluid**. The amniotic fluid prevents the foetus from being squashed and allows the foetus to float freely and have space for its limbs to develop.

Tips for success

Remember that the outer layer of cells of the zygote develops to form the placenta, and the inner layer of cells develops into the foetus.

Spotlight on the test

What is the name of the organ that brings oxygen and nutrients from the mother to the foetus, and at the same time removes waste products and carbon dioxide?   [1]

Check your understanding

20 Use these words to label Figure 2.8.

> vagina   placenta   uterus wall   umbilical cord   amniotic fluid

21 How do the hormones produced by the placenta help the foetus to survive?

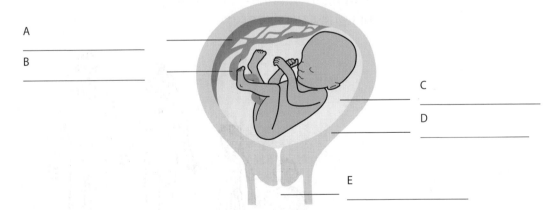

A

B

C

D

E

Figure 2.8 The foetus in the uterus

Health

Conception and pregnancy, growth and development, behaviour and health are all affected by diet, drugs and disease.

Conception and pregnancy

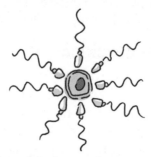
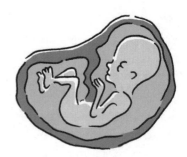

Figure 2.9 Sperm cells, an egg and a foetus

- A diet that is lacking nutrients and is unbalanced can reduce sperm production, stop the menstrual cycle and slow foetal growth, producing smaller babies.
- Alcohol reduces sperm production, smoking tobacco reduces the efficiency of the ovaries, while other non-medical drugs such as cannabis can cause abnormal sperm and stop the menstrual cycle. Alcohol causes nerve damage in the foetus. Smoking tobacco slows foetal growth. Non-medical drugs slow foetal growth, reduce resistance to disease and can make the baby addicted to the drugs.
- Sexually transmitted diseases (STDs) can make females infertile and, in males, can reduce the action of sperm and reduce the chance of fertilisation.
- **Tuberculosis** (TB) can block tubes carrying sperm in males and can make females infertile. TB can kill a foetus. **Malaria** slows the growth of the foetus and can kill it. **Rubella** can cause nerve and heart damage in the foetus, leading to blindness and deafness in babies. STDs can cause blindness, deafness and liver damage in babies.

Figure 2.10 A family unit

Growth and development

- Lack of nutrients can cause poor growth and **marasmus** in babies and **kwashiorkor** in young children. Diets rich in fats and sugars can cause obesity. A balanced diet and a healthy lifestyle with plenty of exercise help to reduce the risk of obesity and develop strong muscles, bones and joints.
- Children may die from sniffing glue.
- A mother suffering from HIV may pass it on to her baby.
- Young people suffering from diseases may grow more slowly or stop growing. Malaria can cause nerve damage which can be fatal.

Behaviour

- A lack of food may produce tiredness. Obese people may have low self-confidence, feel depressed and sleep badly. Hyperactivity may be due to certain food additives.
- Alcohol slows actions, makes movements uncoordinated, and induces sleep or unconsciousness, during which a person may be sick and suffocate in their vomit. Non-medical drugs can cause hallucinations and mental illness.
- Disease causes tiredness and pain. AIDS sufferers may think more slowly, develop a poor memory and have uncoordinated movements.

Health

- If a nutrient in a person's diet is missing then a deficiency disease (see page 10) may develop. A fatty diet may increase the level of cholesterol in the blood and cause circulatory problems.
- Over time, drinking large amounts of alcohol can lead to hepatitis and cirrhosis of the liver. Non-medical drugs can cause brain and heart damage. Smoking tobacco can cause chronic bronchitis and lung cancer.
- Tuberculosis may damage bones, kidneys and the brain. AIDS may weaken bones, and cause kidney and heart disease. Malaria may cause brain, kidney and lung damage.

Figure 2.11 Happy or sad?

Figure 2.12 Healthy

Check your understanding

22 How may alcohol adversely affect a person's life from the foetal stage to becoming an adult?

23 How may a lack of nutrients in the diet affect different stages of life?

 Spotlight on the test

Name *two* diseases caused by smoking. [2]

Tips for success

You will be expected to know about the factors that can affect human health.

3 Cells and organisms

✓ Life processes

There are seven characteristics of life that an organism must show at some stage to be described as living.

Characteristic	Life processes
nutrition	
respiration	
movement	
growth	
excretion	
reproduction	
irritability	

Check your understanding

1 Complete the table of characteristics of life with the following life processes:

> getting rid of wastes   taking in or making food
> changing position   having young   increasing in size
> sensing changes around them   releasing energy from food

Plants and life processes

- Plants make their own food in their leaves by photosynthesis in the daytime (see pages 3–4).
- Plants change position or move as they grow (for example, a climbing bean plant). They are sensitive to features in the environment such as light and will grow towards them.
- Wastes are stored in their leaves and released from the plant when the leaves fall (for example, deciduous trees). Some fully grown plants produce flowers once and die, while others such as trees continue to grow and reproduce for years. Plants respire both day and night.

Animals and life processes

- Animals take in food from plants or other animals. They respire all the time, are sensitive to their surroundings and often respond by moving.
- They grow into adults and reproduce once in their lifetime (for example, insects) or may reproduce regularly over a number of years until near the end of their lives (for example, mammals). Most animals excrete periodically during a day.

Microorganisms

A microorganism is an organism made from a body with only one cell. Microorganisms are important in many ways.

● Decomposition: as bacteria and fungi break down the dead bodies of once living things they release minerals into the soil for plants to use.

● Food production: yeast (a fungus) is used to make bread; bacteria are used to make yoghurt and vinegar.

● Disease: some bacteria cause typhoid, cholera and tuberculosis. Some Protista cause malaria and sleeping sickness. Viruses cause diseases such as measles and chicken pox.

Louis Pasteur and microorganisms

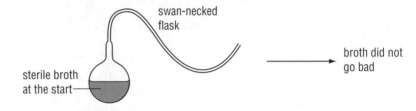

Figure 3.1 Swan-necked flask and broth

● Before scientist Louis Pasteur began his work, people used to believe that living things developed from non-living things to make food go bad.

● Pasteur put some boiled broth in a swan-necked flask and found it did not go bad. When he tipped the broth into the bend or broke open the flask the broth went bad. He reasoned that microorganisms in the air had settled in the neck of the flask and had not managed to reach the broth.

● Pasteur developed 'pasteurisation', a process of stopping liquids from going bad quickly, and discovered that microorganisms caused disease.

Check your understanding

2 Draw a table with these headings and use it to compare the life processes of plants and animals.

Characteristic	Life process in plants	Life process in animals

3 Look at Figure 3.1. Where did the microorganisms settle in the flask?

4 What previous knowledge might have made Pasteur think that boiled broth would not contain living things?

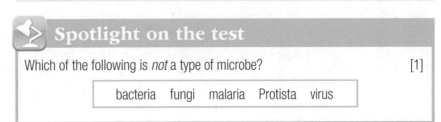

Spotlight on the test

Which of the following is *not* a type of microbe?   [1]

bacteria   fungi   malaria   Protista   virus

From cells to organisms

The **cell** is the basic unit or building block of life. Organisms are made from a large group of cells.

The basic parts of an animal cell

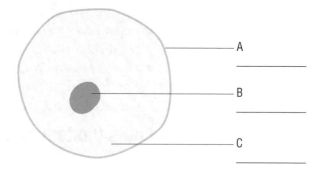

A —

B —

C —

Figure 3.2 A typical animal cell

- A cell has a control centre called the **nucleus**.
- A watery jelly called **cytoplasm** surrounds the nucleus and is the place where most life processes take place.
- The **cell membrane** is a thin sheet of material which surrounds the cytoplasm. It lets food and oxygen pass in and carbon dioxide pass out. It can prevent some harmful substances entering the cell.

The parts of a plant cell

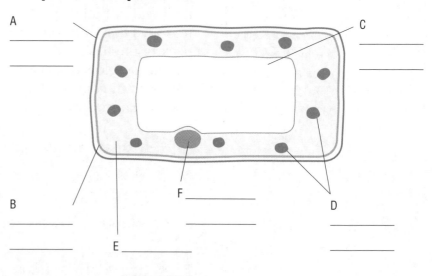

A

C

B

F

D

E

Figure 3.3 A typical plant cell

- A plant cell has the basic parts shown in Figure 3.2, but has some extra features, as Figure 3.3 shows.
- The **cell wall** is made from cellulose, a tough material which supports the cell.
- The **chloroplasts** in the cytoplasm contain chlorophyll which traps energy from sunlight for photosynthesis.
- The **vacuole** contains cell sap which contains sugars, salts and water. The water provides further support to the cell.

Structure and function of common cells

Cell and its structure	Function
muscle cell: spindle shaped, capable of movement	if muscle cells are arranged in layers at right angles they can move food in the oesophagus, stomach and intestine
nerve cell: cell body with a long fibre	electrical signals pass along the fibre, generating messages for the brain and muscles
ciliated epithelium cells: have a surface covered with tiny moving hairs called cilia	cells line the windpipe and the cilia move dust-filled mucus away from the lungs
root hair cell: cell with long thin probing extension into the soil	cells form near the root tip to take up water from the soil

Three common cell types

Tips for success

Make sure you are able to recognise and explain the functions of a range of specialised cells.

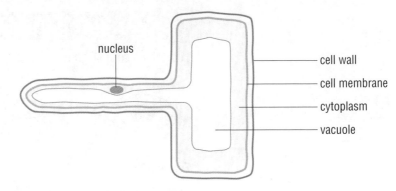

Figure 3.4

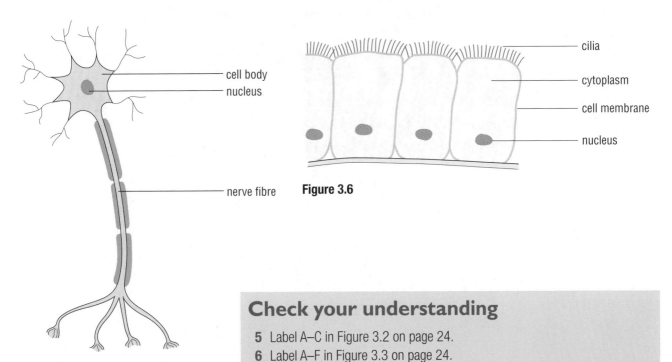

Figure 3.5

Figure 3.6

Check your understanding

5 Label A–C in Figure 3.2 on page 24.
6 Label A–F in Figure 3.3 on page 24.
7 Identify the cells in Figures 3.4–3.6 by giving each diagram a title.

Cells, tissues, organs and organisms

- **Cells** of one type are arranged into groups called **tissues**. The tissue performs a task in the life of the organism. A tissue of ciliated epithelium cells, called the ciliated epithelium, moves dust away from the lungs.
- Groups of **tissues** join together to form an **organ**. The stomach has tissues of smooth muscle cells, tissues of mucus-secreting cells and cells that secrete enzymes for the digestion of protein in food.
- Groups of **organs** which perform a particular task form a **system**. The stomach is joined to the small intestine and other organs to form the digestive system.
- A group of organ systems which together perform all the life processes form an **organism**.

▷ Spotlight on the test

Look at the cell in Figure 3.7.

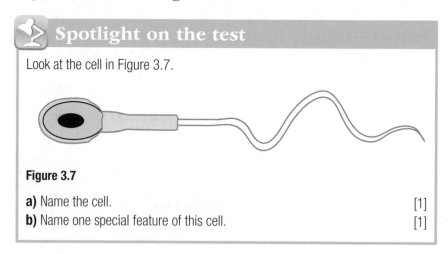

Figure 3.7

a) Name the cell. [1]

b) Name one special feature of this cell. [1]

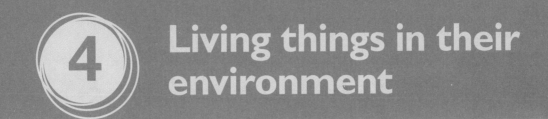

Living things in their environment

 ## Adapting to habitats

- A **habitat** is the place where a plant or an animal lives.
- Each habitat has special conditions and the plants and animals that live there are especially suited or adapted to live in those conditions.

Habitat and conditions	Adaptations of plants	Adaptations of animals
rainforest: hot and humid, sunlit top, dark on floor	trees have their main leafy branches at the top of the trunk to receive light; plants on the floor can make food with low light or grow up the sides of trees to take them into brighter light	monkeys are adapted to balancing on branches, jumping and climbing to find food in the tree tops; frogs have suckers on their feet for tree climbing
desert: hot in day, cold at night, short period of rainfall onto well-drained soil	plants such as a cactus have thick waxy skins to prevent water loss and long roots to find water deep in soil; some plants grow flowers and set seed quickly after rain, then survive as seeds in dry periods	many animals make or find burrows to hide from the heat and cold; they conserve water in their bodies by releasing small amounts of urine; their mouth parts are adapted for feeding on tough plants; camels have wide feet to prevent them sinking in the sand
river	plants take in carbon dioxide dissolved in water; also take in minerals dissolved in water through stem and leaves; roots for holding plant in mud; weak floppy stems as water supports the plant	fish take in oxygen dissolved in water through their gills; have streamlined shape for moving through a much denser substance than air; insects have flatter bodies so water flows more easily over them as they grip rocks; leeches have suckers to hold on in currents

> **Tips for success**
>
> Make sure you can describe a range of adaptations of plants and animals to various habitats.

Adaptations to changes in habitats

The conditions in habitats change daily and seasonally, and living things are adapted to these changes.

- Daily changes: most flowers open in daytime to attract insects for pollination, and close at night for protection from the cold or nocturnal feeders. Some flowers open and produce scent at night to attract moths for pollination. Most birds have eyes that can only see well in daylight. Owls can see well in the dark. Bats use echolocation to find food at night.
- Seasonal changes: many trees shed leaves in autumn or in the dry season to prevent them losing water when frozen ground or lack of rain reduces water for the roots. Many animals such as deer grow thicker coats in winter to keep warm. Some animals hibernate.

Studying habitats

- Plants: mapping a small area is done using a quadrat; finding how plant spacies change down a hillside or from the edge of a wood to the middle is done using a line transect.
- Animals: a Tullgren funnel is used to collect animals from soil and leaf litter; pitfall traps are used to catch small animals moving along the ground; a sheet, beater and pooter are used to collect small animals on branches; a sweep net is used to collect small animals from tall grass.

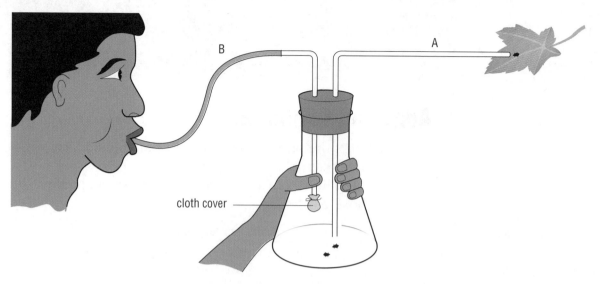

Figure 4.1

Check your understanding

1 Make a table using the headings below to show how the plants and animals in the countryside or a city park near you change during the year, season by season.

Season	Plants	Animals

2 Give the apparatus in Figure 4.1 a title. Annotate the diagram to explain how it works.

Spotlight on the test

Hedgehogs hibernate during winter months. Give one reason why they do this. [1]

Food, energy and populations

Food, the energy food contains and populations of living things are all linked together through feeding.

The food chain and energy flow

- A **food chain** shows the path of food and the energy it contains through a series of organisms as one feeds on another.
- The energy of almost all food chains comes from the Sun, so it is usually omitted from the food chain diagram. The diagram usually begins with the organism that converts the energy in sunlight into the food – a plant.
- Figure 4.2 shows a food chain and the ecological words that are linked to each organism in it. Note that animals that feed on both plants and animals are called omnivores.

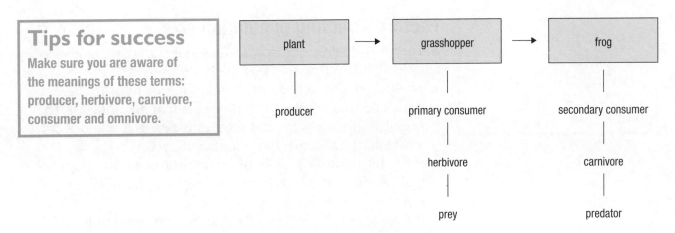

Tips for success

Make sure you are aware of
the meanings of these terms:
producer, herbivore, carnivore,
consumer and omnivore.

Figure 4.2 A food chain and its terms

The food web and energy flow

The **food chains** in a habitat link together to form a **food web**. A food web in a forest habitat is shown in Figure 4.3.

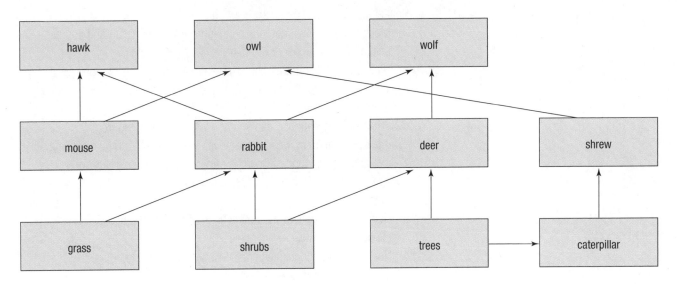

Figure 4.3 A food web in a forest habitat

The role of decomposers

● When a plant or an animal dies its body is broken down by **decomposers**.
● Decomposers may be animals, such as insects and worms, and microorganisms, such as bacteria and fungi.
● Decomposers feed on dead material until all that remains are the minerals that settle into the soil to be used by plants which take part in new food chains. Their feeding process allows the minerals to be recycled.

Tips for success

Make sure you are aware of the factors that can affect the birth rate and death rate of a population.

Factors affecting population size

- The **population** of a species in a habitat is the number of individuals of that species which are present.
- The two major factors which affect population size are the birth rate (the number of individuals that are added to the population each year) and the death rate (the number of individuals that leave the population each year).
- If the birth rate is greater than the death rate, the population will increase. If the birth rate is less than the death rate, the population will decrease.
- If a species has plenty of food and good weather for raising young, its birth rate will increase. A lack of food and poor weather can reduce birth rates.
- If a species is attacked by a large number of predators or if its habitat is destroyed, its death rate will increase.

Check your understanding

3 Snakes eat frogs.
  a) Make a copy of the food chain in Figure 4.2 on page 29 and add a snake to it.
  b) Which way did you draw the arrow? What does the arrow represent?
  c) What further information can you give about the snake? Is it a producer or consumer, carnivore or herbivore, prey or predator?
  d) What extra information can you now provide about the frog?
  e) If the population of grasshoppers was infected with a fatal disease, how would the population of frogs be affected? Explain your answer.

4 Look at the food web in Figure 4.3 on page 29. Imagine that the amount of grass increased.
  a) How would this affect the populations of mice and rabbits? Explain your answer.
  b) How would the populations of the hawk and the owl change? Explain your answer.

Spotlight on the test

Look at the food chain below.

grass → rabbit → fox

Use the following words to describe each of the species:

carnivore   herbivore   producer

a) Grass is a _____.
b) The rabbit is a _____.
c) The fox is a _____.   [1]

Humans in the environment

Once humans left their hunter-gathering ways and began farming and building towns, they affected their surroundings in many negative ways. In recent years they have come to affect the environment in some positive ways but there is still much to do to keep the Earth a habitable place for living things, including ourselves.

The positive and negative influences of humans

	Positive influences of humans	Negative influences of humans
Energy	Using renewable energy: wind, water and solar sources Introducing energy-saving schemes such as switching off lights when not in use Using methane from wastes as a fuel	Using fossil fuels causes an increase in carbon dioxide in the air and may increase global warming Soot, smog and acid rain are produced
Pollution	Reducing carbon dioxide emissions by using catalytic converters Reducing the use of CFCs Limiting the use of pesticides and fertilisers Recycling to reduce the need for landfill sites	As above Use of CFCs damages the ozone layer; wastes from factories, excessive use of fertilisers and pesticides in farming causes water pollution Oil spillages pollute seas
Materials	Recycling to reduce the need to extract more from the Earth and to conserve for the future	Mining can destroy habitats
Habitats	Creating national parks for wildlife prevents habitat destruction	Towns, cities and farms destroy habitats

Energy sources

There are two types of energy sources: **renewable** and **non-renewable**.

Non-renewable resources

- Fossil fuels: coal, oil and natural gas (methane). These have many negative influences (see *Energy* on pages 72–76).
- Nuclear fuel: radioactive materials such as uranium.
  A positive feature is that it does not produce carbon dioxide, so has no link with the possible risk of global warming.
  A negative feature is that a leak of radioactive material from a power station or from stored waste can cause widespread environmental damage.

Renewable resources

- Fuel wood: must be harvested in such a way that woodlands can grow back.
- Solar energy: solar panels trap heat from the Sun's rays; solar cells convert light energy into electrical energy.
- Wave energy: energy in the up and down motion of waves is converted into electrical energy.
- Hydro-electric power station: energy in moving water released from a dam is converted into electrical energy. A negative feature is habitat destruction to create the water reservoir for the power station.
- Wind energy: movement of air turns turbines and is converted to electrical energy. A negative feature is that the blades can kill birds.

Check your understanding

5 After reading about the positive and negative influences of humans on the environment, draw a set of weighing scales and show how you think the balance should look.

6 How does the building of a city affect the environment?

 Spotlight on the test

Explain why it is important to limit the use of fertilisers in farming.　　　　[1]

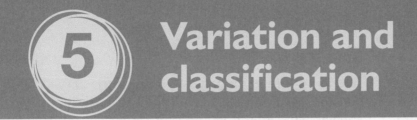

Variation and classification

☑ Groups, species and keys

Living things are grouped or classified according to the features they possess. In this **classification system** the two major groups are the plant kingdom and the animal kingdom.

Major groups of plants and animals

- The **plant kingdom** is divided into five large subgroups.
- The **animal kingdom** is divided into animals without an internal skeleton and backbone (**invertebrates**) and animals with an internal skeleton and backbone (**vertebrates**).

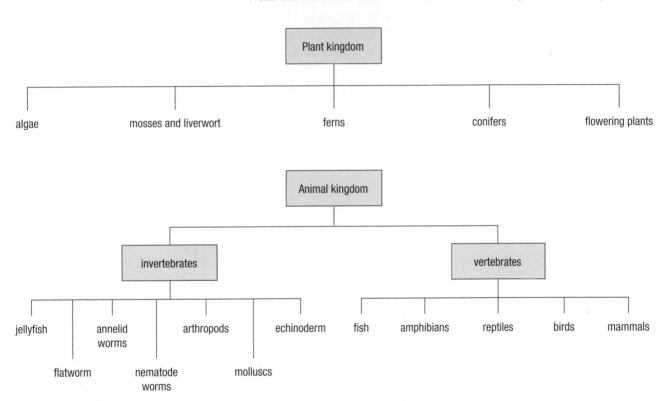

Figure 5.1 Major groups of the plant and animal kingdoms

- The members of each major group have one or more features in common. For example, jellyfish have soft jelly-like bodies and arthropods have a hard skeleton on the outside of their bodies.
- The major groups are also divided into groups. The **arthropod** group, for example, is divided into the insect group, spider group (arachnids), crab and lobster group (crustaceans) and centipede and millipede group. Members of each group share a major feature, for example all insects have six legs.

Species

The **kingdom** is the largest group in the classification system and the **species** is the smallest group. All the living things in a species have a large number of similarities and the males and females of the same species can breed new individuals. Males and females of most species cannot breed with the males and females of other species.

Keys

A **key** is a series of statements that can be used with observations on a living thing to help to identify it. The statements may be set out in a spider key, where you read from the top centre and move down the statements on the 'legs' until the organism is identified.

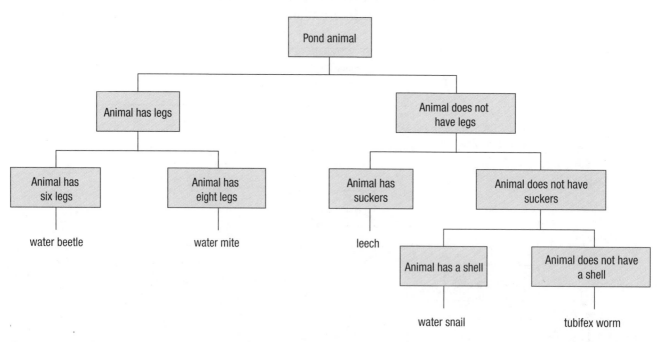

Figure 5.2 A spider key

Check your understanding

1 Is an insect a vertebrate? Explain your answer.
2 Is a kingdom in a species or a species in a kingdom?
3 Make a key about three animals from these statements and names:

> no legs   wings   butterfly   no wings   snake   legs
> two wings   monkey   four wings

 Spotlight on the test

Give a reason why the classification system of plants and animals is so important.   [1]

Inheritance

An inheritance is something which you receive from another person. In science, inheritance means receiving features, also called characteristics, from parents.

Genetic material

- The **nucleus** of a cell contains threads called **chromosomes** on which are zones called **genes**. Genes are made up from **DNA**.
- The DNA in a gene carries an instruction for the body to make a particular characteristic such as brown eyes or black hair.
- Chromosomes are arranged in pairs in the nucleus so there is also a pair of genes for each characteristic. When **gametes**, such as those in eggs and sperm, are produced, they receive just one of each pair of chromosomes.
- When the gametes fuse in fertilisation the **zygote** receives one half of its chromosomes from each parent, and so has pairs of chromosomes made up of one from the mother and one from the father.
- The two genes from the two parents may have different instructions for certain characteristics. The way the pair of genes pass on their instructions produces an individual that varies in some ways from both parents.

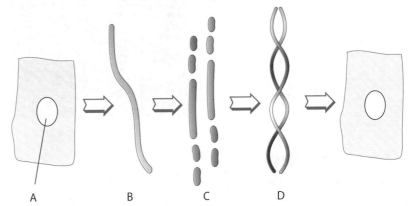

A   B   C   D

Figure 5.3 Genetic material

Selective breeding

- The individuals in a species vary due to the way they inherit their genes. Selective breeding is used to produce varieties in a species which have particular useful features.
- For example, a crop plant species may have some individuals which produce larger grains than others. These are then bred together so that they produce a variety with even larger grains. The new variety may have a stalk that is too weak to hold the larger head of grain, so a strong-stalked variety is brought into the breeding programme to make a plant with large grains and a strong stalk.

Figure 5.4 Charles Darwin (1809–82) commemorated in this statue

Charles Darwin

- Charles Darwin travelled on a survey ship called HMS *Beagle* that was making a world tour. He studied the bodies of the living things he saw and then tried to explain how their features varied.
- In the Galapagos Islands, Darwin found finches with different shaped beaks, each adapted to a particular food such as insects or seeds. He explained these observations by setting out a theory called **natural selection**.
- Using this theory Darwin explained that all the finches may have developed or **evolved** from an earlier finch. Some had slightly pointed beaks, which were adapted to feeding on insects, while others had slightly blunt beaks, which were adapted to feeding on seeds. Over time, the birds with the pointier or blunter beaks survived better than the others and left more offspring with their characteristics. After many generations two new species would have formed.

Figure 5.5 A domestic sheep   **Figure 5.6** A wild sheep

Spotlight on the test

Look at the pictures above of the domesticated sheep used in modern farming and the wild sheep. Suggest *one* feature that has been selected for the modern domesticated sheep. [1]

Check your understanding

4 a) Label the parts of Figure 5.3 on page 35.
  b) Draw a pair of chromosomes and genes in the fertilised cell and give it a title.

5 What kind of features would you use to selectively breed a plant to grow and produce a crop in an area with periods of dry weather?

6 Which of the following features of scientific enquiry do you think Darwin used in developing his theory? Circle your answers.

> made experiments over a long time
> used creative thought
> used a wide range of apparatus including thermometers and measuring cylinders
> made careful observations
> used first-hand experience

States of matter

☑ Particle theory

- The **particle theory** states that materials are made from particles that are so tiny they cannot be seen with the naked eye.
- These particles are atoms and molecules.

Particles in solids, liquids and gases

- The particles in solids are held together by strong forces. The particles cannot change position and this gives solids a fixed shape and a definite volume, and makes them hard to squash. The particles, however, can vibrate a little in their fixed positions.
- The particles in liquids are held together by weaker forces, so they can slide over each other. This gives liquids a definite volume and makes them hard to squash, but allows them to flow and take up the shape of any container into which they are poured.
- The particles in gases have very small forces between them, so the particles can move freely in all directions. This allows gases to flow easily, be squashed easily, change their volume and take up the shape of their containers.

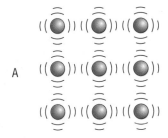

Figure 6.1 Particles in the three states of matter

Changes of state

- Changes due to raising the temperature: heating a solid to a certain temperature (the melting point) makes it change into a liquid, in a process called **melting**. Heating a liquid to a certain temperature (the boiling point) makes it change into a gas in a process called **boiling**.
- Evaporation: a liquid can change into a gas in a process called **evaporation**. Evaporation occurs over a range of temperatures below the liquid's boiling point.
- Changes due to lowering the temperature: if a gas is cooled it may reach a temperature at which it **condenses** and changes into a liquid. If a liquid is cooled down, it may reach a temperature at which it **freezes** (the freezing point) and turns into a solid.

Gas pressure

- A gas is made of millions of quickly moving particles which bounce off the walls of their container.
- The pushing force they generate as they bounce exerts a pressure on the walls of the container. If the gas is heated, the particles receive more energy, move faster, bounce more frequently and raise the gas pressure. Cooling lowers the gas pressure.
- If the temperature is kept constant and the volume is reduced, the particles bounce more frequently on the smaller area of the container walls and the gas pressure is increased.

Diffusion

- **Diffusion** is a process in which one substance spreads through another. It can occur when two or more liquids or two or more gases meet.
- The particles in diffusing liquids move around each other and the particles in diffusing gases bounce off each other as they spread out. Diffusion is faster in gases because gas particles move faster than particles in a liquid.

Check your understanding

1 To which state of matter do the groups of particles in Figure 6.1 on page 37 belong? Write your answer next to A, B and C.
2 A reversible change is one that can be reversed. What is the reverse of:
   a) melting
   b) boiling
   c) freezing
   d) evaporating?
3 What would happen to the pressure of a gas if the volume of its container was increased?
4 How do you think raising the temperature affects the process of diffusion? Explain your answer.

Spotlight on the test

Complete the sentences below by inserting one of the following words:

| gas | liquid | solid |

a) Particles in a _____ are held together by weak forces.
b) Particles in a _____ are held together by very weak forces.
c) Particles in a _____ are held together by strong forces. [1]

Material properties

Everyday materials

There are many everyday materials. The most frequently used materials are plastic, metals, fabrics, wood, pottery, glass, paper, cardboard, brick and concrete.

Physical properties

The physical properties of a material are the features it possesses that can be investigated by observation or by simple experiments which do not involve chemical reactions.

Tips for success

Look around you at a range of everyday objects. Think about the materials they are made from. What properties do they have that make them suitable for their functions?

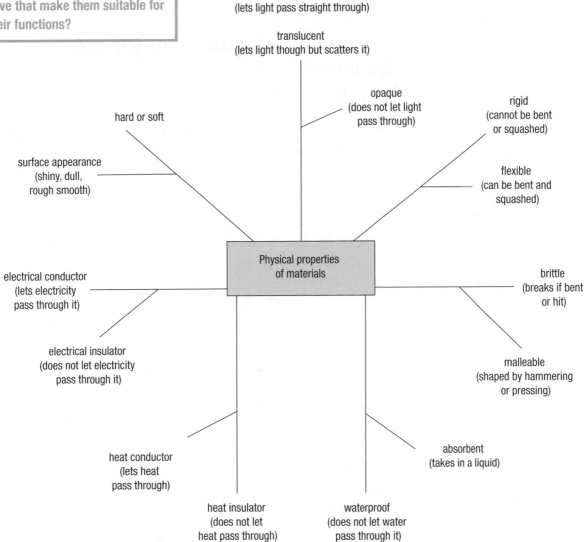

Figure 7.1 Physical properties of materials

The physical properties of metals and non-metals

Elements are substances made from one type of atom. They can be classified as metals (for example, iron, copper, silver and gold) and non-metals (for example, carbon, sulfur, oxygen and chlorine). The table below shows the physical properties of metals and non-metals.

Property	Metal	Non-metal
state at room temperature	solid (one is a liquid)	solid, liquid or gas
density	generally high	generally low
surface	shiny	dull
melting point	generally high	generally low
boiling point	generally high	generally low
effect of hammering	shaped without breaking	breaks easily
magnetic	a few examples	no examples
conduction of heat	good	very poor
conduction of electricity	good	very poor (one conductor)

The chemical properties of metals and non-metals

- The chemical properties of a substance are the ways in which it behaves in chemical reactions with other substances.
- In these reactions the substance is changed into another substance. For example, if iron and sulfur are heated together the two elements take part in a chemical reaction that produces the compound called iron sulfide.
- The reaction between a metal and oxygen forms a metal oxide (for example, calcium oxide). The metal oxide is called a **base** because it can take part in a chemical reaction with an acid and neutralise it. Metal oxides that dissolve in water are called **alkalis**.
- The reaction of non-metals with oxygen forms non-metal oxides (for example, sulfur dioxide). Most non-metal oxides dissolve in water to form **acids**.

Tips for success

Make sure you are familiar with the names of a variety of elements, both metals and non-metals.

Check your understanding

1 Describe the physical properties of a plastic ruler.
2 Compare the physical properties and chemical properties of iron and oxygen.

Spotlight on the test

Sort the following elements into metals and non-metals.

| chlorine | copper | helium | iron | oxygen | silver | sodium | sulfur |

Metals: _____

Non-metals: _____ [1]

Atoms

- An **element** is a substance which is made from just one type of atom.
- The structure of the atom gives the element its physical and chemical properties.

Rutherford and the atom

- Ernest Rutherford knew from the work of another scientist called Thomson that atoms had negatively charged electrons. He also knew that an atom did not have a negative charge so there must be something else present to cancel out the negative charge.
- Rutherford was involved in the discovery of large positively charged particles called α-particles (α is the Greek letter alpha) and he used them to investigate atomic structure. He set up a thin sheet of gold surrounded by a screen that could detect α-particles, as shown in Figure 7.2.

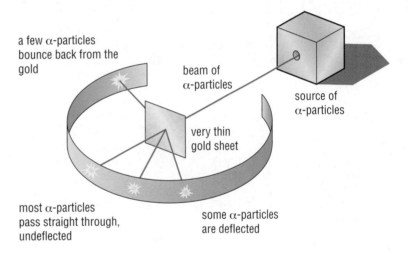

a few α-particles
bounce back from the
gold

beam of
α-particles

source of
α-particles

very thin
gold sheet

most α-particles
pass straight through,
undeflected

some α-particles
are deflected

Figure 7.2 Rutherford's α-particle experiment

- α-particles were fired from the source onto the gold sheet. The screen showed that most α-particles passed straight through the sheet but some had been deflected to other parts of the screen.
- Rutherford explained these results by saying that most of the structure of an atom was empty space, which the α-particles had passed straight through, but at the centre of an atom was a positively charged nucleus which deflected some of the α-particles.
- From this he described the atom as having a central positively charged **nucleus** surrounded by a cloud of negatively charged **electrons**.
- Rutherford later discovered that the atomic nucleus contained **protons** and his work stimulated another scientist to discover that the nucleus could also contain **neutrons**.

The first 20 elements of the periodic table

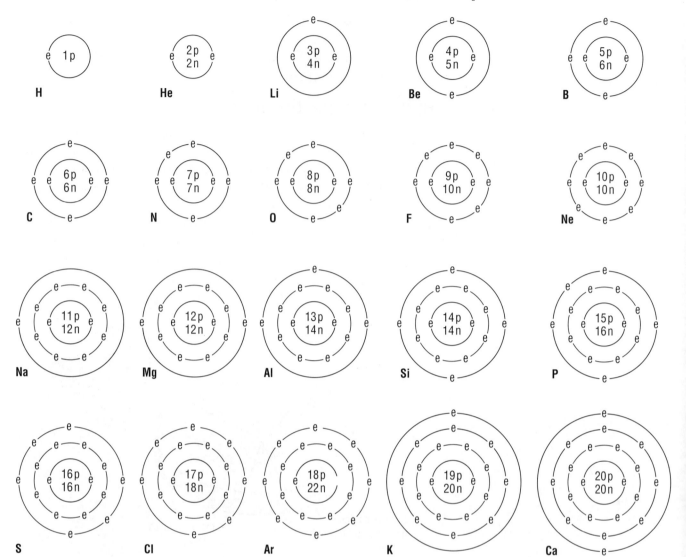

Figure 7.3 The atomic structure of the first 20 elements of the periodic table

Check your understanding

3 What previous knowledge did Rutherford have to help him work out the structure of the atom?

4 How did Rutherford interpret the patterns he saw on the screen in his α-particle experiment?

5 What did Rutherford discover when he refined his method?

6 Next to each symbol in Figure 7.3, write down the name of the element.

Tips for success

Familiarise yourself with the periodic table. You should be able to recognise the names of the first 20 elements.

 Spotlight on the test

How many different types of atoms will a block of copper contain? [1]

The periodic table

As more and more elements were discovered in the nineteenth century, scientists set out to arrange them in order to make their study easier. The result of this work was the **periodic table**.

Scientists and the periodic table

- John Dalton tried to put the elements in order using a measurement called atomic weight.
- Johann Wolfgang Döbereiner found he could group elements into groups of three, called triads, based on their atomic weights.
- John Newlands arranged the elements in order of their atomic weights, starting with the element with the lowest weight. He found that elements eight places apart had similar properties and from this developed his 'law of octaves'.
- Dalton had assumed that the atoms of each element joined with just one atom of another element in a chemical reaction, but over the years other scientists discovered that the atoms of some elements could join with two, three or more atoms.
- Dmitri Mendeleev used all this evidence and developed Newlands' work on grouping to produce the periodic table. When the elements are arranged in rows there is a rise and fall in the number of atoms the elements will combine with.

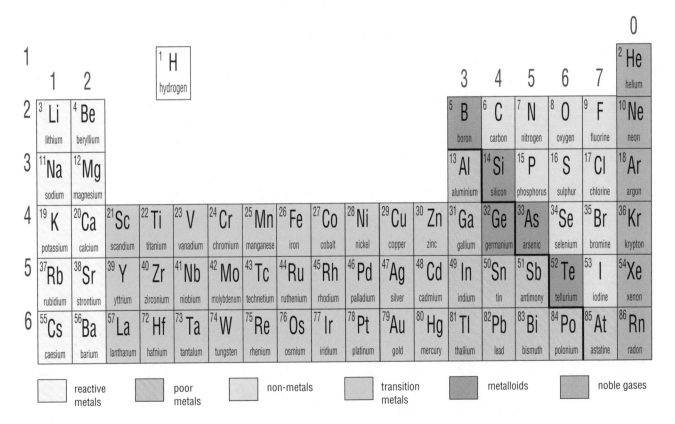

Figure 7.4 Parts of the periodic table

Trends in groups

- Group 1 – the **alkali metals**: going down the column the density generally increases, the melting and boiling points generally decrease, the metals become softer and their reactivity in chemical reactions increases.
- Group 2 – the **alkaline earth metals**: going down the first three elements in this column the densities, melting points and boiling points decrease but their reactivity in chemical reactions increases.
- Group 7 – the **halogens**: going down the column the melting and boiling points increase but their reactivity in chemical reactions decreases.

Trends in periods

Moving along each period from Group 2 to Group 5, the elements change from being metals to being non-metals.

Check your understanding

7 What previous research did Döbereiner and Newlands use in their work?

8 What previous work did Mendeleev use in setting up the periodic table?

9 On a black-and-white copy of the periodic table, colour in the alkali metals in blue, the alkaline earth metals in green and the halogens in red.

Spotlight on the test

Sort the following elements into alkali metals, alkaline earth metals and halogens.

| bromine   calcium   chlorine   magnesium   potassium   sodium |

Alkali metals: _____

Alkaline earth metals: _____

Halogens: _____ [1]

Elements, compounds and mixtures

- An **element** is a substance which is made from just one type of atom. Each element has its own particular properties.
- A **compound** is a substance that is formed from two or more elements which have joined together as the result of a chemical reaction.
- The physical and chemical properties of a compound are different from the physical and chemical properties of the elements from which it is formed.
- The elements in the compound can be separated from each other when the compound takes part in other chemical reactions.

Compound	Description	Use
copper oxide	black powder	colouring pottery
carbon dioxide	colourless gas	taken in by plants for photosynthesis
calcium hydroxide	white solid	used to make lime water
potassium chloride	white crystals	fertiliser
copper sulfate	white powder or blue crystals	fungicide
calcium carbonate	white solid	making snail and egg shells

- A **mixture** is made up of two or more substances. Each substance is spread out through the other substance or substances. The substances can be either elements or compounds, or both elements and compounds.
- The substances in a mixture keep their own physical and chemical properties. For example, if iron powder is mixed with sulfur it retains its magnetic properties and can be separated from the mixture using a magnet.
- The substances in a mixture can be separated by physical processes such as filtering, boiling and condensing.

Separation mixtures

- An insoluble solid can be separated from a liquid by **filtration**. The holes in the filter paper are large enough to let the liquid through, but small enough to hold back the solid particles.
- A soluble substance and its solvent can be separated by **distillation**. The solvent boils and turns into a gas when it is heated. It is then directed into a cooler region where it condenses to form a liquid again. The soluble substance cannot boil away and so is left behind as a solid.

Tips for success

Make sure you can explain the differences between elements, compounds and mixtures.

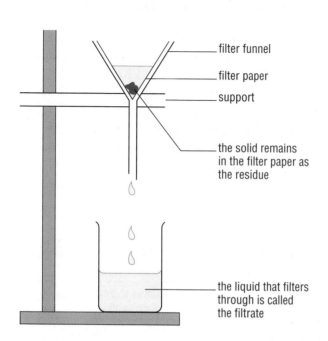

Figure 7.5 Filtration with a filter funnel

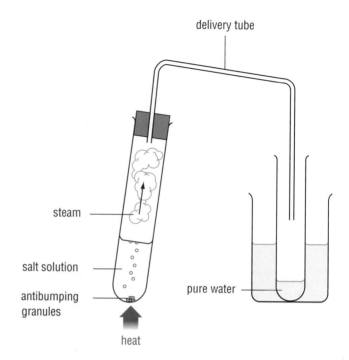

Figure 7.6 Simple distillation

Check your understanding

10 Fill in this table by writing in the letter of the appropriate answer:

A spread out   **D** changed when combined
B physical processes   **E** chemically combined
C retained   **F** chemical reactions

	Compound	Mixture
Elements are …		
Properties of the elements are …		
Elements are separated by …		

11 Mark with an X where filtration occurs in Figure 7.5 and mark with a Y where condensation occurs in Figure 7.6.

Spotlight on the test

State *two* methods of separating mixtures.

_____ and _____ [1]

8 Material changes

☑ Acids and alkalis

- Acids and alkalis are liquids.
- Acids have a sour taste while alkalis have a soapy feel as they react with fat on the surface of the skin. These are *not* tests to try in the laboratory or at home!
- Both acids and alkalis can be corrosive substances and can burn the skin severely.

Indicators

An **indicator** is a liquid which produces a particular colour if it is mixed with an acid or alkali, as the following table shows.

Indicator	Colour in acid	Colour in alkali
red cabbage	changes from purple to red	changes from purple to green
litmus solution*	changes from purple to red	changes from purple to blue
methyl orange	pink	yellow
phenolphthalein	colourless	pink

* Litmus paper is frequently used. Blue litmus paper is used to test for acids. Red litmus paper is used to test for alkalis.

The pH scale

- The letters 'p' and 'H' stand for the 'power of hydrogen'. Hydrogen is found in acids and takes an active part in their chemical reactions.
- The divisions in the pH scale are used to show the strength of acids and alkalis.
- pH 0 to pH 2 indicates a strong acid, pH 3 to pH 6 indicates a weak acid, pH 7 indicates that a solution is neither acid nor alkali and is described as neutral, pH 8 to pH 11 indicates a weak alkali, and pH 12 to pH 14 indicates a strong alkali.
- Universal indicator is a mixture of indicators which gives a range of colours over the pH scale.

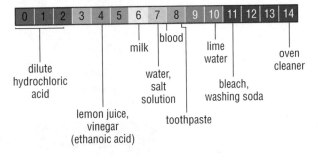

Figure 8.1 The pH scale and universal indicator colours

Neutralisation

- **Neutralisation** occurs when an acid and an alkali take part in a chemical reaction. A salt and water are formed.
- The general word equation for a neutralisation reaction is:

  acid + alkali → salt + water

- Here are some examples from a school science laboratory:

  hydrochloric acid + sodium hydroxide → sodium chloride + water

  sulfuric acid + potassium hydroxide → potassium sulfate + water

  nitric acid + sodium hydroxide → sodium nitrate + water

Applications of neutralisation

- A bee sting is acidic and is neutralised by soap; a wasp sting is alkaline and is neutralised by vinegar.
- Calcium hydroxide is an alkali called lime which is added to soil to neutralise its acidity.
- Sodium hydrogencarbonate dissolves in water to produce an alkaline solution which can be used in neutralisation. The word equation is:

  sodium hydrogen-carbonate + hydrochloric acid → sodium chloride + carbon dioxide + water

- Sodium hydrogencarbonate is used in tablets to cure indigestion when there is too much acid in the stomach. It is used in baking powder, mixed with an acid, and the gas produced makes the cake rise. It is also used in some fire extinguishers; it reacts with an acid and the gas produced pushes water onto the fire and helps extinguish the flame.

Tips for success

Make sure you understand what a neutralisation reaction is and that you can recognise a number of examples of neutralisation reactions.

Check your understanding

1 What colour does blue litmus paper change to if an acid is present?
2 What colour does red litmus paper change to if an alkali is present?
3 What does the pH scale show the following to be:
   a) hydrochloric acid
   b) oven cleaner
   c) milk
   d) water
   e) limewater?

Spotlight on the test

Complete the following word equation for a neutralisation reaction.   [1]

_____ + sodium hydroxide → sodium sulfate + _____

Chemical reactions

Chemical reactions are taking place all round us and inside us, too. Some take in heat (**endothermic reactions**) and some give out heat (**exothermic reactions**). Some of these reactions affect food and other materials and are described as not useful.

Reactions which are not useful

- Some of the reactions which are not useful take place when oxygen reacts with fats and oils in foods. This oxidation reaction makes the foods smell unpleasant and inedible.
- Some manufactured foods have antioxidants added to them to slow down oxidation. Bags of snack foods like corn chips have nitrogen in them instead of oxygen so the oxidation process does not begin until the bag is opened.
- If iron or steel is in damp air, water condenses on the surface and oxygen dissolves in it. The oxygen and iron then take part in an oxidation process in which rust is produced and the metal becomes weaker.
- If very fine powders such as coal dust (in coal mines) and flour (in flour mills) mix with oxygen in the air and become hot they can cause an explosion.

Endothermic reactions

Melting ice
Ice takes in heat from its surroundings when the temperature rises above 0 °C. The ice starts to melt

Sherbet
Sherbet is a popular sweet that is made from citric acid and sodium hydrogencarbonate. When the sweet is put in the mouth it feels cool because it takes in heat from the body and the two compounds react to produce sodium citrate, carbon dioxide and water. (The carbon dioxide gas also makes the sweet 'fizzy')

Endothermic reactions

Cooking
When foods are cooked they take in heat, which allows chemical reactions to take place. These reaction change the texture and taste of the food

Making lime
Limestone (calcium carbonate) is heated to make lime (calcium oxide), which is used to make bricks, plaster, glass and paper

Figure 8.2 Endothermic reactions

> ### Tips for success
> Make sure you can explain what an oxidation reaction is, and why these reactions are not always useful.

Exothermic reactions

Burning
When a substance burns it reacts with oxygen in the air and this reaction releases heat. Fuels such as wood, coal, oil and natural gas are burnt to provide heat for cooking, warming homes and making steam in power stations to generate electricity. Carbon dioxide and water are produced when fuels are burnt

Exothermic reactions

Tips for success
Remember that exothermic reactions give out heat to their surroundings, and endothermic reactions take in heat from their surroundings.

Respiration
Respiration is the process in which energy is released from food (glucose). It takes place in the bodies of all living things. The word equation is

glucose + oxygen → carbon dioxide + water

The energy is used for making the substances needed in the body for growth and in animals it is used by muscles to allow movement

Rusting
When iron and oxygen react in an oxidation process to produce rust, heat is given out. In hand warmers this heat is held in a bag which can be used to warm the body

Figure 8.3 Exothermic reactions

Check your understanding

4 Rusting is speeded up by water and oxygen reaching the metal surface. How can oil prevent rusting?
5 Is melting a physical or a chemical reaction?
6 Oxidation occurs when chemicals react with oxygen. List *five* oxidation reactions featured on pages 49 and 50.

Spotlight on the test

Complete the following word equation. [1]

iron + _____ + _____ → iron oxide (rust)

Reactions of metals

Different metals react at different speeds when they react with other chemicals.

Reaction with oxygen

When heated in oxygen:
- sodium: bursts into flame and burns quickly to form a white powder
- iron: small particles glow and produce yellow sparks
- copper: does not glow or burst into flame but forms a black powder
- gold: does not change.

Reaction with water

- Potassium floats and bursts into flame; hydrogen is produced and potassium hydroxide solution is formed.
- Sodium floats and fizzes; hydrogen is produced and sodium hydroxide solution is formed.
- Calcium sinks and produces bubbles of hydrogen and forms calcium hydroxide solution.
- Magnesium, zinc and iron do not react with liquid water but react with steam in the apparatus shown in Figure 8.4 to produce the metal oxide and hydrogen.
- Copper does not react with water or steam.

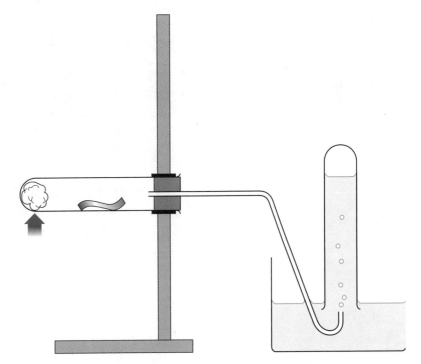

Tips for success

Remember that you can get an indication of how reactive a metal is by observing how vigorously it reacts.

Figure 8.4 Apparatus for investigating the reaction of metals and steam

Reaction with dilute acids

Some metals react violently with dilute acids while others do not react at all. Between these two groups of metals are other metals which react at different rates (see the last column in the table below).

The reactivity series

Metals can be arranged in the order they react with oxygen, water and dilute acids. This classification is called the **reactivity series**.

Metal	Reaction with oxygen	Reaction with water	Reaction with acid
potassium	oxide forms very vigorously	produces hydrogen with cold water	violent reaction
sodium			
calcium		produces hydrogen with steam	rate of reaction decreases down the table
magnesium			
aluminium			
zinc			
iron	oxide forms slowly		
tin	oxide forms without burning	no reaction with water or steam	very slow reaction
copper			no reaction
silver	no reaction		
gold			

Tips for success

Make sure you are familiar with the reactivity series, and why metals are placed in that order.

Check your understanding

7 What compounds do sodium, iron and copper form with oxygen?
8 On Figure 8.4 on page 51, label the metal sample with A. Put B where hydrogen is collected. Label rocksil wool soaked in water with C. Draw an arrow labelled D to show where heat is applied to the apparatus.
9 Why is copper placed above tin in the reactivity series?

Spotlight on the test

Explain why gold is considered to be a very unreactive metal. [1]

Preparing common salts

Common salts can be prepared from a metal or metal carbonate and one of three acids: hydrochloric acid, sulfuric acid and nitric acid.

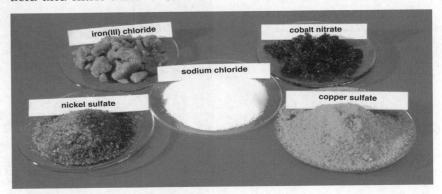

Figure 8.5 Some examples of chloride, sulfate and nitrate salts

Preparing a salt from a metal and an acid

The stages in preparing a salt are:

1 Add the metal to the acid in a flask. Let all the hydrogen bubbles escape before moving to Stage 2.

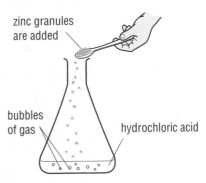

Figure 8.6 Stage 1

2 Pour the contents of the flask into the filter funnel to separate the solid from the liquid.

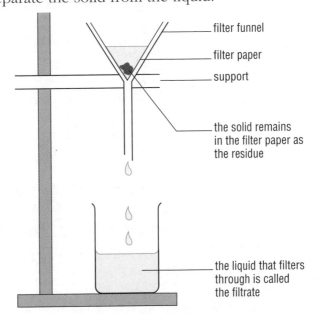

Figure 8.7 Stage 2

● CHEMISTRY

Tips for success

Make sure you are familiar with a range of acids used to make salts, as well as the names of salts they will produce.

3 Pour the liquid into an evaporating dish. Heat it gently until a solid appears.

Figure 8.8 Stage 3

4 Leave the mixture to cool.
5 Filter the mixture again to collect the solid salt.

Preparing a salt from a metal carbonate and an acid

When a metal carbonate is used instead of a metal to make a salt, the same procedure is used.

Check your understanding

10 When a salt forms from an acid what physical process takes place?

11 What process continues to take place as the mixture cools?

12 The stages in the preparation of a metal carbonate are shown below but in the wrong order. Rearrange them to make a flow chart.

   A Stop heating when the solid appears.
   B Filter the carbonate from the mixture.
   C Add carbonate to the acid in a flask
   D Let the evaporating dish cool.
   E Pour the mixture into an evaporating dish and gently heat.
   F Filter the mixture to collect the solid salt.

Spotlight on the test

Complete the following word equation: [1]

calcium carbonate + hydrochloric acid → _____ + _____

\+ _____

54

☑ Rates of reaction

Measuring rates of reaction

1 The rate of some reactions can be found by measuring the reduction in mass of the reactants over a few minutes if a gas is produced and escapes.

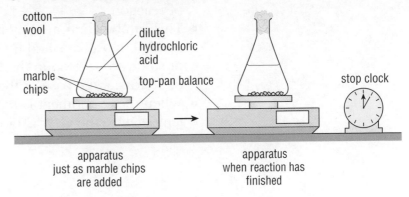

Figure 8.9 Measuring the change of mass of the reactants

2 The rate of reaction can be found by measuring the volume of gas produced during the reaction over a few minutes.

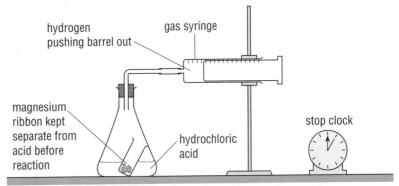

Figure 8.10 Measuring the volume of gas produced in a reaction

Concentration

- The **concentration** of the liquid reactant (the acid) is a measure of how much solute is dissolved in it.
- If there is a small amount of solute in a large volume of solution the concentration is low, but if there is a large amount the concentration is high.
- Higher concentrations increase the rate of reaction. Lower ones decrease the rate of reaction.

Particle size

- A cube with sides 2 cm long has six sides. Each side has a surface area of $2 \times 2 = 4 \, cm^2$. The total surface area of the cube is $6 \times 4 = 24 \, cm^2$.
- If the cube is broken into eight cubes with sides 1 cm long, each of which has six sides, then the surface area of each cube is $6 \times 1 = 6 \, cm^2$. As there are eight of them, the total surface area is now $6 \times 8 = 48 \, cm^2$.

- Breaking into smaller particles further increases the total surface area. As the surface is the place where the reaction takes place, increasing the surface area increases the rate of reaction.

Temperature

- Raising the temperature of the reactants speeds up the rate of reaction.
- This can be confirmed by mixing hydrochloric acid and sodium thiosulfate, and timing how long it takes for a cloud of sulfur to develop in the liquid sufficient to block the view of a cross at the base of the flask.
- When the experiment is repeated with reactants at a higher temperature the view is blocked more quickly.

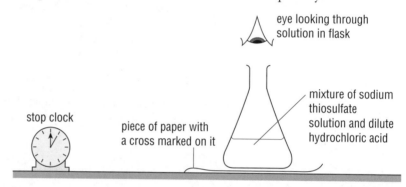

Figure 8.11 Viewing a mixture and measuring the time for the reaction to take place

Catalysts

- A **catalyst** is a substance that speeds up the rate of a reaction but does not change itself.
- Only a small amount of a catalyst is needed to produce a large increase in the reaction rate.
- A catalyst only works on speeding up one reaction; it cannot speed up a number of different reactions.
- Manganese dioxide is a catalyst that is used to speed up the rate at which hydrogen peroxide breaks down into oxygen and water.

Tips for success

Remember that concentration, particle size, temperature and the presence of a catalyst are important factors in determining the rate of a reaction.

Check your understanding

13 State *three* ways to slow down the rate of reaction between a solid and an acid.

14 State *four* ways by which a reaction might be speeded up.

 Spotlight on the test

Suggest *one* way by which the reaction between magnesium and hydrochloric acid could be speeded up. [1]

The Earth

Earth structure, rocks and soils

Earth structure

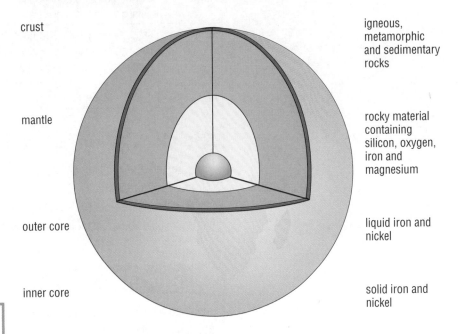

crust	igneous, metamorphic and sedimentary rocks
mantle	rocky material containing silicon, oxygen, iron and magnesium
outer core	liquid iron and nickel
inner core	solid iron and nickel

Figure 9.1 The structure of the Earth

Tips for success

Make sure you can remember the names of the three main rock types, and recognise the features of each of them.

Rocks

There are three types of rock.

Rock type	Rock	Features
igneous: escaped from volcanoes	basalt	black; small crystals due to rapid cooling
	granite	pink, white with black dots; large crystals due to slow cooling
	obsidian	black with glassy structure
sedimentary: produced by weathering and deposition	sandstone	made from grains of sand cemented together
	limestone	made from shells of ancient sea creatures
	chalk	made from tiny shells of ancient planktonic organisms
	shale	made from clay and mud particles
metamorphic: made by squashing and heating in the crust	marble: made from limestone	white, glistens like sugar
	slate: made from shale	grey; breaks to form thin sheets

Soils

There are three main types of soil. Sand, clay and silt are made by the weathering of rocks. Humus forms from the rotting remains of plant and animal bodies.

Soil type	Main component	Drainage	Air content	pH
sandy	sand grains; some humus	very good	good	acidic
clay	clay particles; some humus	poor	low	alkaline
loam	sand, silt, clay and large amounts of humus	good, but humus holds some water for roots to use	very good, humus makes soil crumbs with large air spaces	varied by farmers to match the plants they wish to grow

Check your understanding

1 Complete the labelling in Figure 9.1 on page 57 by adding pointer lines to the correct features.
2 How can the elements in granite also be found in slate?
3 How could you improve the drainage of a soil?

 Spotlight on the test

Jonathan has found a rock containing shells of tiny ancient sea creatures.
a) Name the rock type. [1]
b) Explain how it was formed. [1]

 # Fossils and the age of the Earth

Fossils

- **Fossils** are formed from the bodies of ancient plants and animals that are quickly covered by mud or sand, which prevents decay. Over time, the mud and sand turn to rock and as they do, water passing through them can produce fossils in one of two ways: replacement or petrification.
- Replacement: the body tissues dissolve and are washed away to be replaced with minerals which come out of water and make a rocky shape of the body.
- Petrification: minerals in the water passing through the fossil settle in the tissues and turn them into rock.

Rock layers

- Rocks break down in the process of **weathering** and the tiny fragments are washed away and deposited in layers, often at the mouths of rivers.
- Over time, layers of different rock such as limestone and sandstone build up on each other. The oldest layer is usually found at the bottom and the most recent layer is found at the top.
- Observing how layers of sand and mud are built up today hints that rock formation takes a long time and that the Earth is very old.

The fossil record

If the fossils are set out in the order they are found in the rocks, starting at the bottom with the oldest rocks and moving to the most recent at the top, a fossil record is made. Here is an example of a fossil record. '✓' in the table shows the presence of a particular type of organism.

Layer name	Formation began (mya)*	Flowering plants	Trilobites	Insects	Reptiles
Quarternary	about 2.5	✓		✓	✓
Tertiary	about 65	✓		✓	✓
Cretaceous	about 145	✓		✓	✓
Jurassic	about 200	✓		✓	✓
Triassic	about 251			✓	✓
Permian	about 300		✓	✓	✓
Carboniferous	about 360		✓	✓	✓
Devonian	about 416		✓	✓	
Silurian	about 444		✓		
Ordovician	about 488		✓		
Cambrian	about 542		✓		

* mya = millions of years ago.

Fossils and rocks

- Different layers of rock can have groups of fossils of organisms that lived in different habitats, such as swamps and deserts. This suggests that during the time the different layers were laid down, the climate changed. Up until now this has been a very slow process, again hinting that the Earth is very old.
- The fossil record shows that fossils change over time, in a process called evolution. Recent studies on evolution show that it takes a very long time, another hint that the Earth is very old.

Recent estimates on the age of the Earth

- The age of a rock is found by studying the amounts of **radioactive elements** it contains.
- A radioactive element is one that breaks down or decays to form other elements. Radioactive elements decay at a regular rate, with some taking thousands or even millions of years to decay to half of their original mass.
- A definite age for a rock can be found by comparing the amounts of radioactive elements in it with the amounts of elements they have produced. Using this method the age of the Earth is estimated at 4.6 billion years.

Tips for success

Make sure you are aware of the evidence that points to the Earth being very old.

Check your understanding

4 For each of the following questions you need to give an explanation as part of your answer.
   a) Did a trilobite ever see a flower?
   b) Have insects always fed on flowering plants?
   c) Could a reptile have eaten a trilobite?
5 What features of rocks and fossils hint that the Earth is very old?
6 Why does using radioactive materials give a more reliable estimate of the age of the Earth?

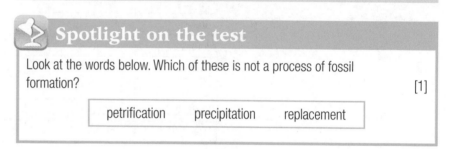

Spotlight on the test

Look at the words below. Which of these is not a process of fossil formation?   [1]

| petrification | precipitation | replacement |

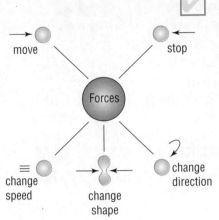

Figure 10.1 Action of forces

☑ Forces

Contact and non-contact forces

- There are two kinds of contact forces: impact forces, when surfaces touch and strain forces, when a force builds up in the object on which a force is acting. An example of a strain force is the force inside a squashed sponge ball. Tension is a strain force that builds up in a stretched elastic band.
- A non-contact force that is exerted by one object on another without the two objects touching. Examples are magnetic, electrostatic and gravitational forces.

Friction

Friction is a contact force that occurs between two surfaces when a push or pull could cause one surface to move over the other.

- Static friction: if a small pushing or pulling force is generated to slide one surface over another a frictional force of similar strength is generated between the surfaces in the opposite direction. Gradually increasing the pushing or pulling force causes a gradual increase in the frictional force to match it. The force that exists between the two surfaces when there is no movement is called static friction.
- Sliding friction: if the pushing or pulling force is further increased, one surface will slide over the other because the static friction cannot stop it. However, the surface cannot slide freely because there is a sliding frictional force which is less than the maximum value of the static frictional force. If the pushing or pulling force ceases, the sliding frictional force stops the two surfaces sliding.

Increasing and reducing friction

All surfaces have tiny projections (bumps) and hollows.

- Reducing friction: a liquid applied to the two surfaces fills the hollows and makes the contact between the surfaces smoother.
- Increasing friction: pressing the surfaces makes the projections and hollows lock together. Making a surface rougher also increases friction.

> **Tips for success**
>
> Make sure you can give examples of contact and non-contact forces.

> **Tips for success**
>
> Make sure you can explain the ways of increasing and reducing friction, and why both can be important.

Air resistance

- When an object moves through the air, the object pushes on the air and the air pushes back with a force called **air resistance**.
- The faster an object moves, the more it pushes on the air and the greater the air resistance pushing back on it.
- Vehicles which are designed to travel very fast through the air, such as a racing car or an aeroplane, have a **streamlined shape**. These shapes have a pointed front (aeroplane) or a wedge shape (cars) and curved body surfaces to make the air flow over the body easily, which reduces the air resistance.

Check your understanding

1 Describe the forces acting in this situation: a car is moving along, it hits a cardboard box and makes it move.

 Spotlight on the test

Sean says that gravity is a contact force. Explain why he is not correct.   [1]

✓ Gravity

Six facts about gravity

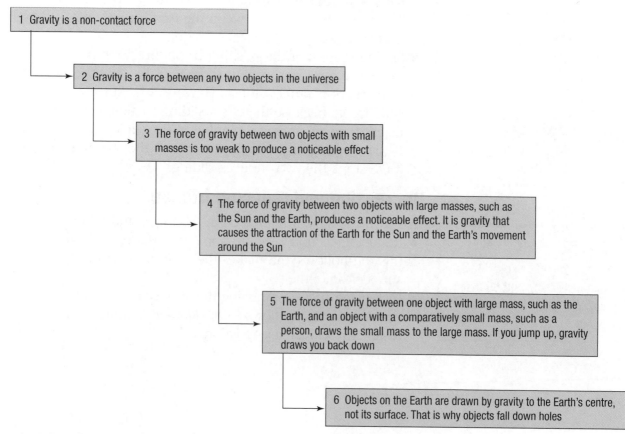

Figure 10.2 All about gravity

Mass and weight

- The **mass** of an object is the amount of matter in it and is measured in grams (g) or kilograms (kg).
- The **weight** of an object is the pull of the Earth's gravity on the object and is measured in newtons (N).
- The pull of the Earth's gravity on an object of mass 1 kg is almost 10 N and for calculations is considered to be 10 N. This means an object with a mass of 3 kg has a weight of $3 \times 10 = 30$ N.

Gravity and weight in other places

- The region in which a non-contact force acts is called a **field**.
- Gravitational field strength is calculated by dividing weight by mass. For example, on Earth an object of weight 10 N has a mass of 1 kg, so the gravitational field strength is 10 N/kg.
- The gravitational field strength of the Moon is six times weaker than that of the Earth. This means the weight of an object on the Moon is one-sixth the weight on Earth. The gravitational field strengths of the other planets in the Solar System also differ from that of the Earth.

Stretching springs

- Objects used to investigate springs are called masses. If a spring is hung vertically and a mass is attached to the end, the weight of the mass pulls down and stretches the spring.
- If the size of the mass is increased, the extension of the spring increases in proportion, up to a point called the **elastic limit**. Up to this limit the removal of the mass allows the spring to return to its original length. Beyond this limit the spring is permanently deformed and cannot return to its original length.

The newton spring balance

- The spring balance is used in laboratory investigations on forces. It contains a spring and a device called a stop which prevents the spring being stretched beyond its elastic limit.
- The spring can be used to measure and compare forces because it stretches in a regular way as forces are applied to it. The amount of stretch is measured on a scale which is calibrated in newtons.

Check your understanding

2 An object has a mass of 9 kg.
   a) What is its weight on the Earth?
   b) What is its weight on the Moon?
   c) What is its weight on Mars? (Mars has a gravitational field strength one-third of that of the Earth.)

⚡ Spotlight on the test

Charlotte would like to measure the mass of her science book.
1 Name the instrument she should use. [1]
2 Name the units of measurement. [1]

✓ Speed

Speed is a measure of the distance covered by a moving object in a certain time.

time

start

finish

distance travelled

Figure 10.3 Measuring speed

Measuring speed with a stopwatch

- A distance is marked out, with a start line and a finish line.
- When an object moves away from the start line, the stopwatch is started. When the moving object passes the finish line, the stopwatch is stopped.
- The distance between the start and finish lines is then divided by the time taken to give a value. The value has units of distance/time, such as metres/second.

Measuring speed with a light gate

- A light gate is often shaped like an upside-down U, with the light source near the tip of one arm and a light-sensitive switch near the tip of the other arm. When a moving object passes between the arms it breaks the beam of light.
- In an investigation to measure speed, a light gate is set up at the start line and the switch is set to start an electronic clock when the beam is broken. A second light gate is set up at the finish line with the switch set to stop the electronic clock when the beam is broken. In the investigation the object passing the start line starts the electronic clock and when it passes the finish line it stops the electronic clock.
- The electronic clock can give a more accurate measure of speed than using a stopwatch.

Distance–time graph

- A distance–time graph is a graph showing the distance travelled by an object in a certain time. It can be used to find the speed of an object.
- The *x* axis (horizontal) of the graph always shows the time, in units such as seconds, minutes or hours. The *y* axis (vertical) of the graph always shows the distance travelled, in units such as centimetres, metres or kilometres.

Tips for success

Remember:

$$speed = \frac{distance}{time}$$

Units of speed should always be given in your answers.

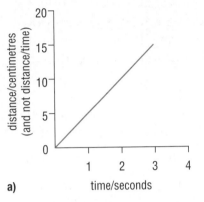
a)

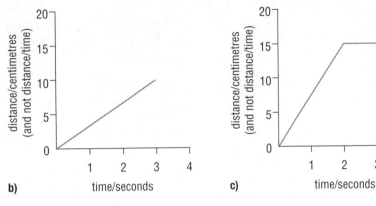
b)   c)

Figure 10.4 Examples of distance–time graphs **a**), **b**) and **c**). Graph **a**) shows an object travelling at a speed of 5 cm/second for 3 seconds.

Check your understanding

3 What is the speed in cm/second of the following objects:
  a) distance travelled = 50 cm, time at first light gate = 0.00 s, time at second light gate = 10.00 s
  b) distance travelled = 100 cm, time at first light gate = 0.00 s, time at second light gate = 20.00 s
  c) distance travelled = 120 cm, time at first light gate = 0.00 s, time at second light gate = 15.00 s
  d) distance travelled = 144 cm, time at first light gate = 0.00 s, time at second light gate = 36.00 s
4 Which is faster, the object in 1a) or in 1c)?
5 Which is slower, the object in 1b) or in 1d)?

 Spotlight on the test

An Olympian runs a 100 metre race in 10 seconds. What is her average speed?   [2]

Pressure

- **Pressure** is the term used to describe a force acting over an area of known size. It can be defined by the equation:

$$\text{Pressure} = \frac{\text{force}}{\text{area}}$$

- The units for pressure are N/cm² or N/m².

Pressure exerted by a solid

- The pressure exerted by a solid object acts on the surface below it. The weight of the object in newtons pushes down on the area in contact with the surface. If the weight of the object is 100 N and the area in contact is 1 m² then the pressure is $\frac{100}{1}$ = 100 N/m².
- Pressure can be reduced by keeping the area of contact the same and reducing the size of the force. For example, if the force is reduced to 50 N, then the pressure is $\frac{50}{1}$ = 50 N/m².
- Pressure can be reduced by increasing the area in contact. For example, if the area of the object in contact with the surface is increased to 2 m², the pressure is $\frac{100}{2}$ = 50 N/m².
- Pressure can be increased by keeping the area in contact the same and increasing the size of force. For example, if the force is increased to 200 N, the pressure is $\frac{200}{1}$ = 200 N/m².
- Pressure can be increased by reducing the area in contact. For example, if the area in contact is reduced to 0.5 m² the pressure is $\frac{100}{0.5}$ = 200 N/m².

Pressure exerted by a liquid

- The particles in a liquid move around and exert pressure not only on the bottom of the container but also on its sides, as Figure 10.5 shows.
- The pressure at the bottom of the liquid is higher than that at the top due to the weight of liquid above it.
- The taller the column of water, the greater the pressure difference between the top and the bottom. This is why a dam wall has a wider stronger base compared with the top, to stand up to the stronger pressure at the bottom of the reservoir.

Tips for success

With all calculations, remember to include units in your answers.

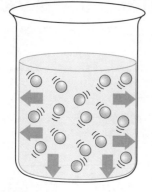

Figure 10.5 Forces exerted by particles in a liquid

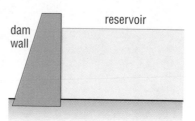

Figure 10.6 Cross-section of a dam wall and reservoir

Pressure exerted by a gas

- A gas contains millions of quickly moving particles. Large numbers of the particles push on the walls of the container and exert a force. This force divided by the area of the container walls produces the gas pressure.
- The pressure can be raised by heating the container and gas. This makes the particles move faster and push on the walls more frequently.
- The pressure can also be raised by decreasing the volume of the container so that the particles push on its walls more frequently.
- The pressure can be reduced by cooling the container and the gas. This makes the particles move more slowly and push on the walls less frequently.
- The pressure can also be reduced by increasing the volume of the container so the particles push on its walls less frequently.

Tips for success

Make sure you can explain the effects of changes in temperature and volume on gas pressure.

Check your understanding

6 Do you exert more pressure on the ground when you sit on it or when you lie down on it? Explain your answer.

7 Figure 10.7 shows a cylinder of water with three pipes in its side. There was a piece of rubber tubing attached to each pipe, with a clip to prevent water escaping, but now the clips have been removed and the water is free to flow out of the pipe. Draw the paths of the jets of water on the diagram and explain your answer.

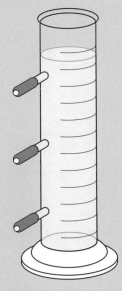

Figure 10.7 Jets of water leaving a measuring cylinder

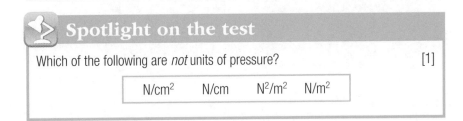

Spotlight on the test

Which of the following are *not* units of pressure? [1]

| N/cm^2 | N/cm | N^2/m^2 | N/m^2 |

☑ Density

- The **density** of a substance is a measure of the amount of matter that is present in a certain volume of it. It can be defined by the equation:

$$\text{density} = \frac{\text{mass}}{\text{volume}}$$

- The units for density are g/cm^3 or kg/m^3.

Determining the density of a solid

- The density of a regular-shaped solid, such as a cube, can be found easily. The mass is found by placing it on a top-pan balance. The volume is found by measuring the length of its sides and multiplying them together.
- For example, if a cube has a mass of 160 g and its sides are 2 cm long then its volume is $2 \times 2 \times 2 = 8\,cm^3$ and its density is $\frac{160}{8} = 20$ g/cm³.
- The density of an irregular-shaped solid, such as a pebble, can be found in the following way.
  - The mass is found by placing it on a top-pan balance.
  - A measuring cylinder is filled up to a certain volume and the volume is recorded as V_1.
  - The pebble is lowered into the water until it is completely covered, as shown in Figure 10.8. This makes the water level rise. The new volume V_2 is then recorded.
  - The volume of the pebble V_p is found by subtracting V_1 from V_2.
  - The density of the pebble is given by mass of pebble/V_p.

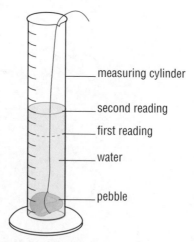

Figure 10.8 Measuring the density of a pebble

Determining the density of a liquid

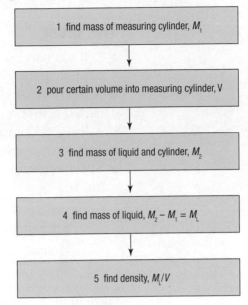

Figure 10.9 Measuring the density of a liquid

Determining the density of a gas

The density of air is determined by the following procedure:

1 The mass of the gas container and air is found by placing it on a top-pan balance. This is M_1.
2 The air is sucked out with a vacuum pump, then is placed back on the balance to find its mass. This is M_2.
3 The volume of the container is measured by opening the container under water. Water replaces the vacuum, so that pouring the water out into a measuring cylinder gives its volume, V.
4 The density is found using the formula:

$$\frac{M_2 \times M_1}{V}$$

> ## Tips for success
>
> Remember:
>
> $$density = \frac{mass}{volume}$$
>
> Remember the units:
>
> $$\frac{g}{cm^3} \text{ or } \frac{kg}{m^3}$$

Check your understanding

8 a) What is the density of a substance that has a mass of 100 g and a volume of 20 cm³?
   b) What is the mass of a substance that has a density of 10 g/cm³ and a volume of 70 cm³?
   c) What is the volume of a substance that has a density of 4 g/cm³ and a mass of 100 g?

9 A pebble displaces 15 cm³ of water when placed in a measuring cylinder and has a mass of 45 g. What is its density?

◈ Spotlight on the test

Which of the following are *not* units of density? [1]

| g/cm³ | kg/m³ | g/cm² | kg/m |

Moments

The lever

- A force can be used to move an object so that it follows a circular path.
- A device that changes the direction in which a force acts is called a **lever**.
- A lever is composed of two arms and a **fulcrum** or pivot.
- The force applied to a lever to do work is called the **effort**.
- The force which resists the action of the effort is called the **load**.

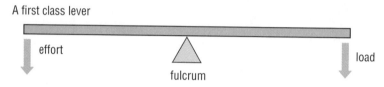

A first class lever

effort

fulcrum

load

Figure 10.10 A lever

Forces and moments

- The turning effect produced by a force around a fulcrum is called the **moment of the force**. The direction of the moment is usually specified as clockwise or anticlockwise.
- In Figure 10.10, the direction of the moment of the load is clockwise and the direction of moment of the effort is anticlockwise.
- The size of the moment is found by multiplying the size of the force by the distance between the point at which the force acts and the fulcrum. In Figure 10.10, these are the distances from the fulcrum to the end of each arm of the lever.
- The moment of a force can be found using this equation:

  moment of a force = force × distance from the fulcrum.
- The moment is measured in newton metres (N m).

> **Tips for success**
>
> Make sure you can remember the equation for calculating moments. Practise using this equation before the test.

The seesaw

- The seesaw is an example of a lever as shown in Figure 10.10.
- For the seesaw to balance and the two arms to be horizontal, the moment of the weight on one arm must be equal to the moment of the weight on the other.

The principle of moments (law of the lever)

When a body is in equilibrium (or balance), the sum of the clockwise moments about any point (such as the fulcrum) equals the sum of the anticlockwise moments about that point.

Check your understanding

10 If the arms of the lever in Figure 10.10 on page 70 were 1 m long and the size of the load was 10 N, what would the moment and direction of the force be?

11 What would the moment and direction of the effort need to be to make the lever balance, given the moment you have calculated in Question 10?

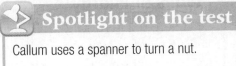

Spotlight on the test

Callum uses a spanner to turn a nut.

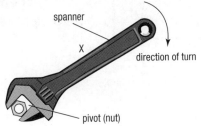

Figure 10.11

The point X is 10 cm from the pivot (nut). He holds the spanner at point X.
He uses a force of 90 N.
Calculate the size of the moment.   [2]

Energy

Energy

- The **law of the conservation of energy** states that energy cannot be made or destroyed, it can only be changed from one form to another.
- Scientists describe energy as a property that something has which allows it to exert a force or to do work.

Types of energy

- There are two kinds of energy: stored or **potential energy** (PE) and movement or **kinetic energy** (KE).
- Potential energy and kinetic energy can be divided further into eight more types (Figure 11.1).

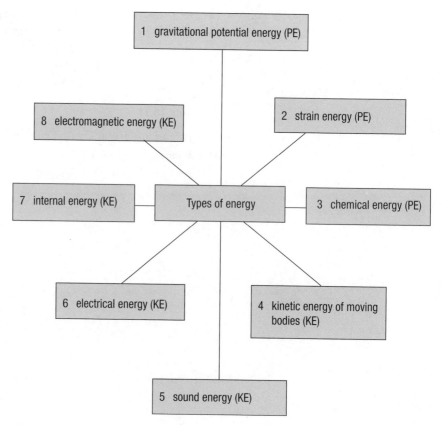

Figure 11.1 Types of energy

Energy transfers

- The transfer of energy can be shown by energy transfer diagrams. The diagrams show the main transfers of energy but it should be remembered that during any energy transfer some energy is always lost as heat. This is not shown on the diagrams unless it is of particular importance, such as in providing heat from a fuel.
- The energy transfer diagram has three parts: energy input and arrow, energy converter or transducer, energy output and arrow

energy input → | transducer | → energy output

- Blowing up a balloon:

kinetic energy → | balloon | → strain energy

The world's energy needs

- There are two kinds of energy resources: **non-renewable energy resources** (NRE) and **renewable energy resources** (RE).
- These can be divided further into eight types of energy resources (Figure 11.2).

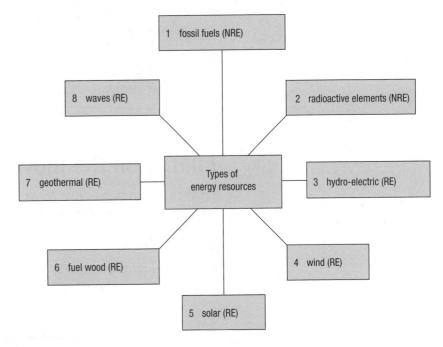

Figure 11.2 Types of energy resources

Non-renewable resources will eventually be used up, so renewable resources should be used alongside them now. The aim should be to switch to renewable forms of energy completely before the non-renewable resources run out.

Check your understanding

1 The following statements are about the eight forms of energy. Write the letter for each statement on Figure 11.1 on page 72 next to the energy it describes:

   A the flow of electricity

   B pulled down by gravity

   C movement of particles inside a substance

   D to and fro vibrations reach the ear

   E energy in stretched or squashed objects

   F two forms are light and heat

   G in links between atoms

   H occurs when objects move

2 In the energy transfer diagram, what has the kinetic energy and what has the strain energy?

3 The following statements are about eight energy resources. Write the letter for each statement on Figure 11.2 on page 73 next to the energy resource it describes:

   A grows on trees

   B from up and down motion of sea water

   C coal, oil, gas

   D uses panels for heat and cells for electricity

   E uranium is the fuel

   F moving air turns turbines

   G moving water turns turbines

   H from heat inside the Earth

Spotlight on the test

List *three* renewable energy resources. [1]

✓ Energy on the move

Heat or thermal energy moves in three ways: by **conduction**, by **convection** and by **radiation**.

Conduction

● In conduction the heat energy passes from one particle in a substance to the next one it touches (Figure 11.3).

heat

Figure 11.3 Conduction of heat

● Conduction can occur easily in solids where the particles remain in one place to pass on the heat, less easily in liquids where the particles move about, and hardly at all in gases where the particles are too far apart to touch frequently enough to pass on heat.

- Conduction does not take place in the vacuum of outer space.
- Materials can be divided into heat conductors and heat insulators. Metals are the best heat conductors because they have freely moving electrons which can carry the heat quickly to cooler electrons and atoms. Air is a good insulator and is trapped in woollen clothes to reduce heat loss from the body.

Convection

In convection, the particles that receive the heat move and take it with them into a cooler region of the substance (Figure 11.4).

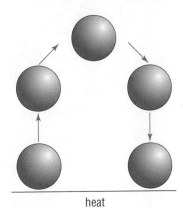

heat

Figure 11.4 Convection

- Convection only occurs in liquids and gases.
- It does not occur in solids, because the atoms do not move, nor in a vacuum such as outer space where there is an absence of particles.

Radiation

- Heat energy can travel as electromagnetic waves by radiation (Figure 11.5).
- These waves are in the infrared region of the electromagnetic spectrum and can pass through the air or a vacuum such as outer space because particles are not involved in their transfer.

heat

Figure 11.5 Radiation wave

Tips for success

Make sure you are clear about the differences between conduction, convection and radiation.

- The surface of a material affects the amount of heat it radiates or absorbs. A black surface radiates and absorbs the most amount of heat in a given time. A light shiny surface, such as polished metal, radiates and absorbs the least amount of heat in a given time.

Evaporation

- The particles in a liquid have different amounts of energy. The higher the energy, the faster they move.
- The fastest moving particles near the liquid surface move so fast that they break through the surface, separate from other liquid particles and become a gas (Figure 11.6).

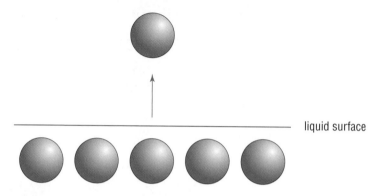

liquid surface

Figure 11.6 Path of a particle in an evaporating liquid

- This process is called evaporation and occurs below the boiling point of the liquid. The gas produced when water evaporates is called **water vapour**.
- When these hotter high-energy particles leave the liquid there is less heat energy in the liquid and its temperature drops.

Check your understanding

4 When a pan of soup is heated, the three kinds of heat transfer and evaporation take place. Describe why and where they occur, starting at the bottom of the pan.

Spotlight on the test

Use the words *particle/s* and *kinetic energy* to explain why a beaker of hot water cools down.   [3]

12 The Earth and beyond

 ## The Earth in space

Lights in the sky

- The Sun and the other stars are sources of light.
- The Moon, the planets and comets reflect light from the Sun towards the Earth.

The rotation of the Earth

- The Earth **rotates** on its **axis** – an imaginary line or rod that runs through the centre of the Earth from the north pole to the south pole. The Earth turns round or rotates once on its axis in 24 hours (one day).
- When a place on the surface of the Earth is turned away from the Sun it is night-time; later, as the Earth continues to turn, that place is turned towards the Sun and it becomes daytime.
- As the Earth turns anticlockwise, as viewed from above the north pole, the Sun always rises over the eastern horizon and sets below the western horizon. In between it seems to rise in the sky in the morning and to sink in the afternoon.
- At night, stars and planets also seem to move across the sky from east to west. This is due to the rotation of the Earth.

The Earth in its orbit

- The Earth's path around the Sun is called its **orbit**.
- The axis of the Earth is not vertical but is at an angle of about 23° to the perpendicular. The Earth's axis always points in the same direction during the Earth's orbit and this produces changes in the amount of sunlight and heat that some places receive.
- These periods in which a place has a certain amount of light and heat (which also affects the weather) are called **seasons**. The seasons in each hemisphere are related to the way the Earth is tilting with respect to the Sun, as shown in Figure 12.1 on page 78.

The path of the Sun in the sky

- The rotating Earth makes the Sun appear to follow a path in the sky from east to west. As the Earth moves in its orbit this path changes (Figure 12.2 on page 78).
- When a **hemisphere** is tilting towards the Sun in the summer time, sunrise is at its earliest, the Sun rises highest at midday and sunset is at its latest. In winter, sunrise is at its latest, at midday the Sun rises to a much lower height and sunset is at its earliest.

> **Tips for success**
>
> Make sure you can explain night and day in terms of the rotation of the Earth.

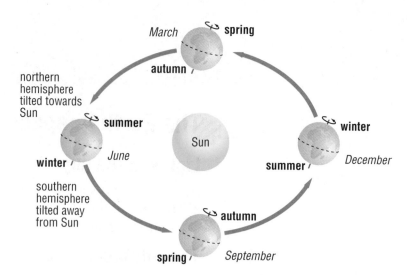

Figure 12.1 The orbit and seasons

● From Figure 12.2 you can also see that the positions of sunrise and sunset are much further to the north in the northern hemisphere in summer compared to winter.

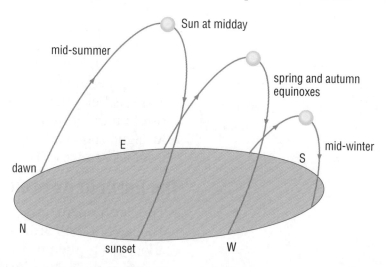

Figure 12.2 The change in the Sun's path with the seasons in the northern hemisphere

Tips for success

Make sure you can explain the seasons in terms of the path of the Sun.

Check your understanding

1 An asteroid is formed from a piece of rock that may have come from a planet. Is it a light source or a reflector of light?
2 If the Earth was tilted at a greater angle than it is, how do you think summers and winters might change?
3 How do the positions of sunrise and sunset change from winter to summer in the southern hemisphere?

Spotlight on the test

The Moon, the planets and comets do not produce their own light, yet they can be seen from Earth. Explain why this is the case. [1]

☑ # The Solar System

The Sun and the planets

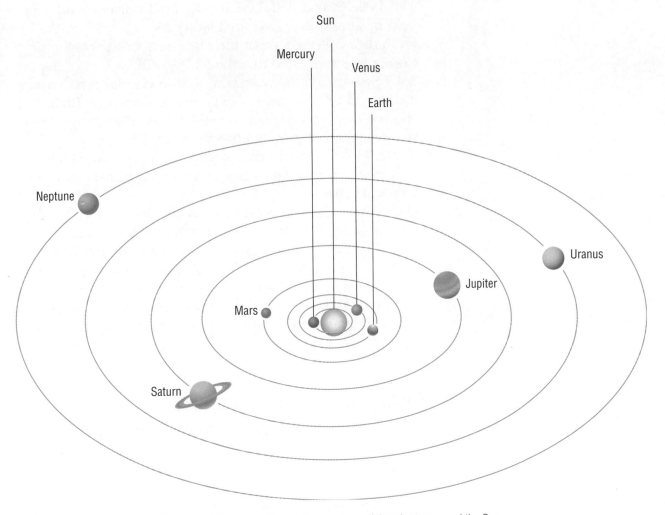

Figure 12.3 The Solar System (not to scale), showing the relative positions of the planets around the Sun

The movement of the planets

The distances of the planets from the Sun, their rotation times and their orbit times are shown in the table.

Planet	Distance from Sun, million km (approx.)	Rotation time			Orbit time, days
		Days	**Hours**	**Minutes**	
Mercury	58	58	15	30	88
Venus	108	243	0	0	224
Earth	150		23	56	365
Mars	228		24	37	686
Jupiter	778		9	50	4332
Saturn	1427		10	14	10759
Uranus	2871		10	49	30707
Neptune	4497	6	15	48	90777

> **Tips for success**
>
> Make sure you can explain the difference between rotation time and orbit time.

Studies on the Solar System

- The ancient Greeks believed that all the objects in the sky were set in crystal spheres which moved around the Earth. This belief was called the Earth Centred Universe and was held by almost everyone for 1300 years.
- From data collected about the movement of the planets, Copernicus suggested that the Sun was the centre of the Solar System and the planets moved in circular orbits around it. Data on the movements of comets collected by Brahe showed that crystal spheres could not exist because comets seemed to smash through them.
- Kepler looked at more data on the movement of planets collected by Brahe and suggested that the planets moved in elliptical orbits.
- Galileo used his telescope to show that Jupiter had moons that moved around it, and not around the Earth, indicating that other objects also might not move around the Earth.
- Newton made calculations on the movement of objects in the Solar System and showed that this movement was due to the force of gravity. All these observations were made on the first six planets in the Solar System, which everyone could see with the naked eye. In 1781 Herschel discovered Uranus and in 1846 Galle discovered Neptune, using telescopes.

Check your understanding

4 Which is the nearest neighbour to Earth – Venus or Mars? Explain your answer.

5 Which are the fastest and slowest spinning planets in the Solar System?

6 What two pieces of evidence did Brahe's data provide that suggested the ancient Greeks were wrong?

7 a) Who provided evidence that all the objects in the sky did not move around the Earth?

  b) What apparatus did he use?

  c) Who else used this apparatus and what did they discover?

 ### Spotlight on the test

Look at the table about the movement of the planets on page 79.

1 Name the planet with the fastest orbit time.   [1]

2 Name the planet with the slowest rotation time.   [1]

13 Sound

✓ The properties of sound

Sound is made when an object **vibrates**. This to and fro movement of the object is transferred to the particles in the air around it. The air particles next to the vibrating object push on the air particles next to them and they to push on air particles further away, and so on. Once a particle has pushed, it swings back to receive another push from the vibrating object and other particles around it. This swinging of the particles produces regions of high and low pressure which move away from the object and create the sound wave.

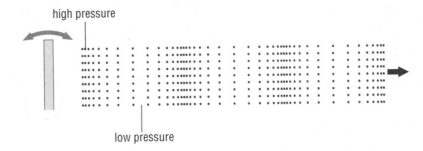

Figure 13.1 Particles and pressure changes

If a microphone is attached to an **oscilloscope** the vibrations of the particles are converted into electrical signals which can be displayed on a screen to produce a wave you can see.

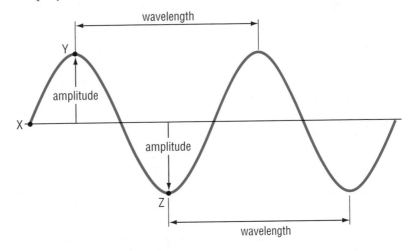

Figure 13.2 The features of a sound wave

The loudness of a sound

- The loudness of the sound is related to the amount of movement of the vibrating object.
- If the object moves only a small distance to and fro it makes a sound wave with a small **amplitude** and the sound will be quiet.

Check your understanding

1 Does sound travel through the vacuum of space? Explain your answer.
2 An object is set vibrating and the sound wave it produces is recorded on an oscilloscope, as shown in Figure 13.3.
   a) What has happened to the vibrating object and the loudness and pitch of the sound in part 2?
   b) What has happened to the vibrating object and the loudness and pitch of the sound in part 3?
   c) What has happened to the vibrating object and the loudness and pitch of the sound in part 4?

- If the object moves a large distance to and fro it makes a sound wave with a large amplitude and the sound will be loud.

The pitch of a sound

- An object makes a certain number of vibrations in a second.
- The vibrations generate the same number of sound waves in a second and this produces the pitch of a sound.
- The number of sound waves produced per second is called the **frequency**. This is measured in hertz, with the unit symbol Hz.
- A low-pitched sound has lower frequency waves with longer **wavelengths** than a high-pitched sound, which has higher frequency waves with shorter wavelengths.

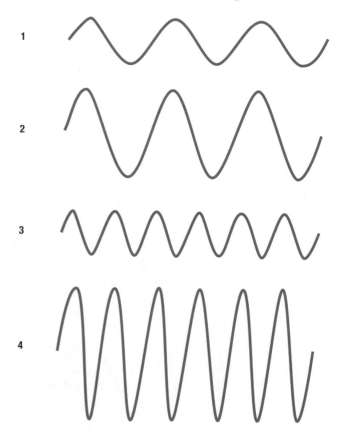

Figure 13.3 Sound waves as displayed on an oscilloscope screen

Spotlight on the test

Josh has produced a diagram of a sound wave using an oscilloscope. He notices that it has a large amplitude and a low frequency. Describe the sound. [1]

 **Light**

 ## Light on the move

Light and shadows

- Light rays travel in straight lines. When light shines on an **opaque** object, rays striking its surface are stopped and there is an absence of light behind the object and this forms the **shadow**.
- The shape of the shadow is not identical to the shape of the object. For example, if the light rays shine down on the object they make the shadow shorter than the object. If the light rays shine from one side of the object they make the shadow taller than the object.
- The size of the shadow on a screen can be increased by moving the object towards the light source and decreased by moving the object towards the screen.

> **Tips for success**
>
> Make sure you can explain how the size of a shadow can be affected by altering the distance between object and light source.

Reflection

- **Non-luminous** objects are seen because they reflect light from a light source into the eyes.
- Most objects have a slightly rough surface. This makes light rays scatter in all directions when they strike the surface, so they do not produce reflections in their surfaces.
- Some objects, such as a mirror, have a very smooth surface. When a light ray strikes it, the ray is reflected as shown in Figure 14.1. This orderly reflection of the light rays, following the **law of reflection**, produces images in the mirror. These images come from the light rays that travel from objects in front of the mirror.

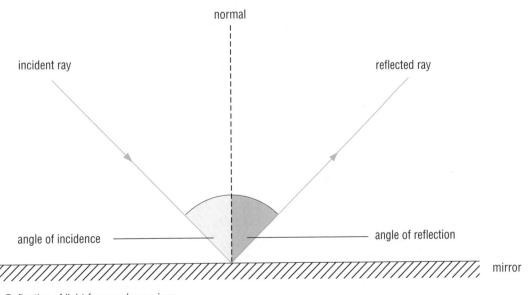

Figure 14.1 Reflection of light from a plane mirror

Refraction

- If a light ray is shone through the air so that it strikes the surface of a **transparent** material such as glass or water at right angles (perpendicular to the surface; see 'normal' in Figure 14.1 on page 83) it passes straight through.
- If the light ray travelling from a less dense material (air) strikes the more dense materials (glass) at another angle the light ray is 'bent' or **refracted**, as shown in Figure 14.2. The **angle of refraction** is less than the **angle of incidence**.

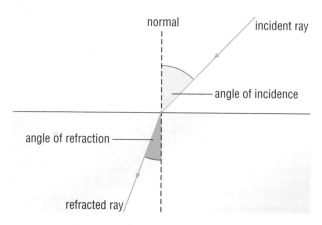

Figure 14.2 Refraction between air and glass

- Figure 14.2 shows how light moving from a less dense material (air) to a more dense material (glass) is refracted.
- When light moves from a more dense material (water) into a less dense material (air) the angle of refraction it makes is greater than the angle of incidence. This makes objects on the bottom of a swimming pool appear closer to the surface than they really are.

Check your understanding

1 How does the shadow of an object change when:
   a) the light source shining on one side is raised above it then lowered again
   b) the object is moved towards a screen and then back towards the light source?
2 If the angle of incidence of an incident ray is 48° what is the angle of reflection?
3 When a light ray shines from air into glass at an angle of incidence of 30° will the angle of refraction be a) the same, b) greater, or c) less?

 Spotlight on the test

Explain how non-luminous objects are seen.   [1]

☑ Colour

The dispersion of white light

A **prism** used to disperse light is a triangular block of glass or plastic. If a ray of sunlight is shone through a prism at certain angles of incidence and its path is stopped with a screen a **spectrum** of light is produced (Figure 14.3).

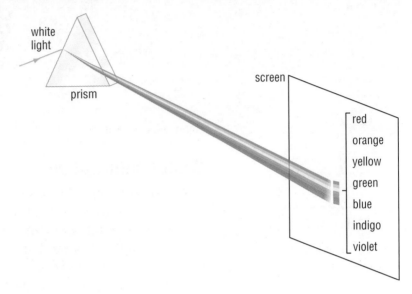

Figure 14.3 Dispersion of white light by a prism

How rainbows are seen

- When a ray of sunlight from behind an observer shines onto a raindrop, the light is dispersed at the surface of the raindrop to form a spectrum on the opposite inner surface of the drop.
- This is then reflected back through the raindrop to a surface below the point of dispersion where it is refracted again and passes to the eye of an observer.

Colour addition

- The **primary colours** of light are red, green and blue. They can be added together by shining their light onto a white screen, as shown in Figure 14.4 on page 86.
- When any two primary colours are added together they produce a **secondary colour**. When all three primary colours of light are added together they produce white light.

Tips for success

Remember how secondary colours and white light can be produced.

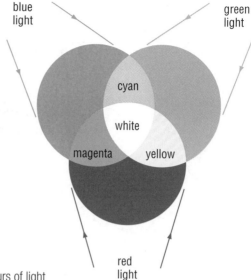

Figure 14.4 Adding colours of light

Colour subtraction

- When white light shines on a painted surface some of the light is absorbed and subtracted from the white light and some of it is reflected and seen by an observer.
- For example, if white light shines on yellow paint all the colours in the spectrum are absorbed except yellow. The other colours are subtracted from the white light, leaving yellow to be reflected and seen by the observer.

> ## Tips for success
>
> Make sure you can explain colour subtraction in terms of reflection and absorption.

Check your understanding

4 Use the information in the section 'How rainbows are seen' on page 85 to construct a diagram of how an observer sees a rainbow.
5 What secondary colours are produced by mixing the following:
   a) blue and green light
   b) blue and red light
   c) green and red light?
6 Magenta, cyan and yellow are used to make coloured paints. What colour of paint is made when:
   a) yellow and magenta are mixed together
   b) cyan and magenta are mixed together
   c) yellow and cyan are mixed together?
7 a) Describe the appearance of a surface which absorbs all the colours in white light.
   b) Explain the appearance of this surface in terms of subtracting the primary colours.

Spotlight on the test

Adele shines a red light and a blue light onto a white table tennis ball. It no longer looks white.
1 What colour does it appear? [1]
2 What colour light would need to be added to make it appear white again? [1]

Magnetism

All about magnets

The properties of magnets

- Iron, cobalt and nickel are the three elements that have magnetic properties. Steel is an alloy (a mixture) of iron and carbon and also has magnetic properties.
- A **magnetic material** is one that is attracted to a magnet and can be made into a magnet.
- Inside a magnetic material are many tiny regions called **domains** which behave like microscopic magnets.
- In an unmagnetised material the domains are arranged at random, pointing in all directions. When the material is magnetised all the domains point in the same direction.
- The regions of greatest magnetic force are near the ends of the magnet. These regions are called the **poles**.
- When a magnet is allowed to move freely, one pole turns towards the Earth's magnetic north pole and is called the north pole of the magnet, and the other pole turns towards the Earth's magnetic south pole and is called the south pole of the magnet.
- When two magnets are brought together their **similar poles repel** each other and their **different poles attract** each other.
- You can compare the strengths of magnets by finding out how many paper clips can hang in a chain from them or how many sheets of card can be placed between the magnet and a paper clip before the paper clip falls away.

Magnetic field pattern

- The **magnetic field** is the region around a magnet where the pull of the magnetic force acts on magnetic materials.
- The magnetic field can be shown by placing a bar magnet on a table, covering it with a sheet of paper, sprinkling iron filings on the paper, and then tapping it. The iron filings then arrange themselves into lines showing the lines of force in the magnetic field.

> **Tips for success**
>
> Make sure you can explain why magnets repel and attract using the terms 'north pole' and 'south pole'.

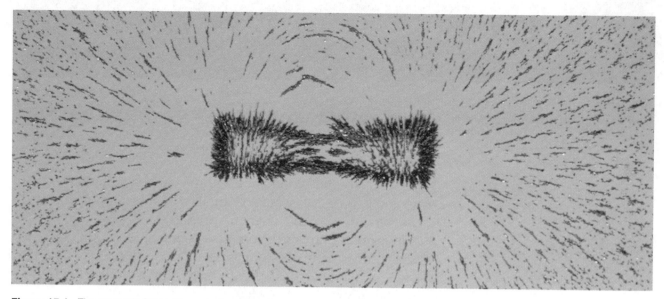

Figure 15.1 The magnetic field pattern of a bar magnet

The electromagnet

- When an electric current runs through a wire it generates a magnetic field around it.
- An **electromagnet** is made by wrapping wire around an iron bar so that the wire makes a coil.
- When an electric current passes through the coil the magnetic field from the coil turns the iron bar into a magnet that can be used to lift things up.
- When the current is switched off the magnetic field in the coil is destroyed, the iron loses its magnetism and the object it was lifting falls off.

Tips for success

Make sure you can identify the components of an electromagnet, and describe how it works.

Check your understanding

1 A paper clip sticks to the north pole of a magnet. What does this tell you about the material the paper clip is made from?
2 If a second similar paper clip is brought near the first, it sticks to it. What has happened to the first paper clip?
3 When the two paper clips are removed from the magnet they no longer stick together. Why?
4 What happens to two magnets:
   a) when their north and south poles are brought together
   b) when their two north poles are brought together?
   Explain each answer.

Spotlight on the test

1 Steel cans can be easily separated from aluminium cans using a magnet. Explain why. [1]
2 What is the major advantage of using electromagnets compared to bar magnets? [1]

Electricity

✓ Electrostatics

The atom and electric charge

- An **atom** has a **nucleus** containing positively charged **protons** surrounded by a cloud of negatively charged **electrons**.
- In an atom, the number of protons and electrons is the same so their charges cancel each other out and the atom is neutral.

Insulators and conductors

- An **insulator** is a material that does not let an electric current pass through it. If it is electrically charged (see below) it holds onto the charge. The electric charge does not go anywhere. It is static.
- A **conductor** allows a current of electricity to pass through it. It cannot hold onto electric charges because it conducts them away.

Charging materials

- When some materials are rubbed they can lose electrons. This means that there are then more positive charges than negative ones and the material becomes positively charged.
- When other materials are rubbed they can gain electrons. This means that there are then more negative charges present than positive ones and the material becomes negatively charged.

The behaviour of charged materials

- If two balloons are negatively charged, hung from threads and brought close together they push each other away because like electrical charges repel each other.
- If a positively charged material is brought near a negatively charged balloon the balloon moves towards it because opposite charges attract.

Induced charge

- An induced charge is one that is made in a material by a charged material without the two materials touching.
- For example, if you rub a plastic pen with a cloth the pen will lose electrons and become positively charged. When the pen is held over tiny pieces of paper it induces a negative charge in their surfaces.
- This can be checked by moving the pen very close to the paper. The paper jumps up to the pen because opposite charges attract.

> **Tips for success**
>
> Make sure you are aware of how electrostatics can be useful, but also be aware of the problems it can cause.

Digital sensors

- Charge is stored in devices called **capacitors**. It is released from the capacitor when it is switched into a circuit with a conductor. The first capacitors had metal plates separated by an insulator.
- Today, capacitors are tiny electronic devices which are used in digital sensors to store electrical charge when changes in light, pH or movement are being recorded. The stored charges provide a record of the changes that have taken place and can be transferred to other devices such as screens which display information.

Sparks and lightning

If two oppositely charged surfaces are very highly charged a spark may pass between them. An example is lightning in a storm. Sparks can also be dangerous; a spark could cause an explosion in fuel vapour.

Check your understanding

1 a) If an atom has seven protons and six electrons what charge is it?
  b) How can the charge be reversed?
2 When pieces of paper jump towards the charged pen, what two forces are acting on them?

Spotlight on the test

Two balloons are rubbed with a dry cloth and hung upside down, next to each other. Describe and explain what happens to them. [1]

Current electricity

How components affect current

Component	How it affects current
cell (battery)	generates current when a circuit is complete
wire	conducts current and offers some resistance
lamp	changes some electrical energy to light and heat, offers some resistance to the current
switch	when closed, it conducts current; when open, air (an insulator) prevents current flow

Series and parallel circuits

- In a **series** circuit all the components are in a line. In Figure 16.1a on page 91 the battery and the two lamps are in a line.
- In a **parallel** circuit some components are arranged side by side. In Figure 16.1b on page 91 the two lamps are arranged in parallel.

Figure 16.1 **a**) A series circuit;
b) a parallel circuit. Note: the positive
terminal of the cell is the longer vertical line

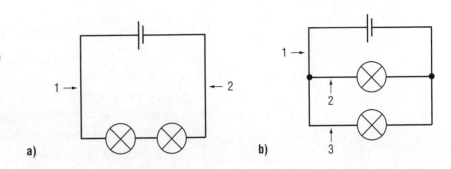

a)

b)

Measuring current

- The rate at which electrons flow through a wire is measured in units called amps, with the symbol A, and is measured using an **ammeter**.
- When an ammeter is placed in a circuit its positive terminal is connected to a wire that leads towards the positive terminal of the cell. It is always connected in series with the component through which the current flow is to be measured.

Division of the current in a parallel circuit

- When the current flowing in a wire reaches two wires with the same resistance in the parallel part of a circuit it splits equally.
- For example, if a current of 4 A is in the single wire each of the parallel wires carries a current of 2 A. When the wires join to make a single wire again the current strength is restored to 4 A.

Voltage

- The voltage is the difference in electrical potential (potential difference) between two points in a circuit and is measured in units called volts, with the symbol V, using a **voltmeter**.
- The voltmeter is connected into the circuit with its positive (red) terminal connected to a wire that leads towards the positive terminal of the cell. The negative (black) terminal must be connected to a wire that leads to the negative terminal of the cell.
- These wires must be attached into the circuit so that the voltmeter is in parallel with the part of the circuit being tested. In Figure 16.2 the voltmeter is measuring the voltage across the lamp.

Tips for success

Make sure you can explain
differences between series and
parallel circuits.

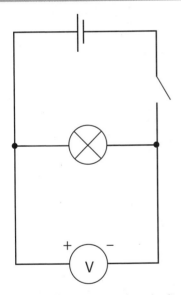

Figure 16.2 A voltmeter in a circuit

Tips for success

Remember that an ammeter
is connected in series with
the component through which
current flow is measured. A
voltmeter is attached in parallel.

Check your understanding

3 Draw a circuit with a cell, a switch and a lamp. Add an ammeter to the circuit to measure the current flow through the lamp.

4 In Figure 16.1a on page 91, if the current at 1 is 3 A what is it at 2?

5 In Figure 16.1b on page 91, if the current at a is 3 A what is it at 2 and 3?

6 Draw a circuit diagram with one cell and three lamps in parallel.

7 What is the difference between a volt and an amp?

Spotlight on the test

Use the components below to draw a circuit with a light that can be switched on and off.

[2]

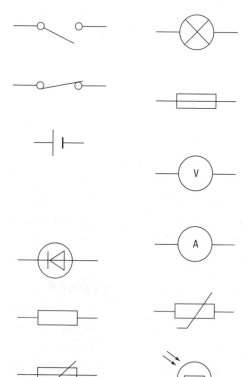

Figure 16.3